TRIGGER WARMING!

Jeffrey R Grimshaw

ISBN: 978-0-6454364-0-2

Front cover design: Jeffrey R Grimshaw

Table of Contents

Chapter XI
APOCALYPSE NOW

Chapter XII
THE STING

Glossary and Abbreviations

Definitions are important especially in the Global Warming arena where words and phrases have had their meanings warped and redefined for political purposes. We have followed the meanings of certain words and phrases as closely to their original intent and definition as possible.

CAGW – **C**atastrophic **A**nthropogenic **G**lobal **W**arming. This is the hypothesis that the Earth is experiencing **W**arming on a **G**lobal scale that is out of the ordinary, the warming is caused primarily by human (**A**nthropogenic) emissions of CO_2 and this warming will be **C**atastrophic for the planet.

Climate Alarmists / Alarmists – The people and institutions that promote the CAGW hypothesis that global warming is man-made and dangerous. Al Gore, Greta Thunberg and Prince Charles are high-profile alarmists.

Climate Change – For the purposes of this book 'Climate Change' means the natural climate change that has occurred for the whole of Earth's history.

Climate Realists / Realists – The people who oppose the CAGW hypothesis. Climate Realists are a broad church of people and groups ranging from those who do not think man-made global warming is a significant threat to those who think that CO_2 cools the atmosphere.

CMIP - Coupled Model Intercomparison Project.

CO_2 – Carbon Dioxide. A molecule of one carbon atom and two oxygen atoms that occurs naturally in the atmosphere as a colourless, odourless gas. It is essential for life on Earth and it provides the carbon that is a major component of all living things.

GHE – Greenhouse Effect.

GHG – Greenhouse Gas.

Global Warming – This term was used for many years by alarmists as a shortening of CAGW. The inference is that Global Warming was happening and caused by mankind. It was later replaced with the more generic term "climate change". We generally stick to the original term of global warming to mean CAGW rather than use climate change. To distinguish CAGW / MMGW from natural climate change we refer to "natural global warming" or "naturogenic global warming".

IPCC – The Intergovernmental Panel on Climate Change.

ISS – International Space Station.

MMGW – Man-made Global Warming – A variation of CAGW.

mw – Megawatt. A unit of power equal to one million watts. Usually used to describe the output of large-scale electricity generators. e.g. "The power station has a capacity of 2000 MW".

mya – Million years ago. e.g. 500mya is 500 million years ago, i.e. half a billion years ago.

MWP – Medieval Warming Period. A period from circa 900 AD to 1400 AD, corresponding roughly to the period known as The Middle Ages. Temperatures during this period were over 1^0C warmer than present. The MWP was followed by The Little Ice Age (1400-1850 AD).

O_2 – Oxygen.

O_3 – Ozone.

ppm – Parts Per Million: A unit of concentration used for small values. One PPM is "one in a million" or 0.0001%.

RWP – Roman Warming Period. A period from circa 250 BC to 450 AD, when Rome was geo-politically dominant. Temperatures during this period were up to 2^0C warmer than present. The RWP was followed by the colder period known as The Dark Ages.

Foreword

"I cannot teach anybody anything. I can only make them think"
SOCRATES

"Why a book debunking man-made global warming?", you may ask. "Isn't the debate over? The science settled?"

Absolutely not.

Science is never "settled" and in the case of global warming, the debate has never actually really taken place. The general public has only seen or heard one side and almost nothing of the many other sides. Yes, there are many opinions and hypotheses on climate, how it functions and what influences it. But the alternative views have been silenced and drowned out by the prevailing narrative that:

1. **The Earth is warming at an unprecedented rate**
2. **CO_2 is the primary factor in determining the temperature of the Earth**
3. **This warming is caused by man-made CO_2**
4. **The warming is catastrophic**
5. **This catastrophic warming can be reversed by mankind**

We have heard many people ask what book they can read that would provide an alternative to the mainstream opinion that there is a deadly invisible threat in the air, that computer models predicted doomsday scenarios of death, suffering and misery, which requires a global emergency to be declared and that the solution to the crisis involves destroying economies, impoverishing people, diminishing individual rights and a global "reset".

There are many alternative books on global warming but few cover all aspects in a short, succinct, easy-to-read manner. So we decided to write our own.

The plan was to complete and publish this book in mid-2020 only to be delayed by an apparent deadly invisible threat in the air, that computer models predicted doomsday scenarios of death, suffering and misery, which requires a global emergency to be declared and that the solution to the crisis involves destroying economies, impoverishing people, diminishing individual rights and a global "reset".

Who would have imagined that?

The policies advocated by our so-called moral betters that are supposed to save the planet are actually inflicting catastrophic damage on the natural environment. They also inflict harm on people and on society, the tone of public debate, the quality of journalism, the credibility of scientific research and education and the integrity of government policy and public administration.

The damage fails a cost-benefit analysis and cannot be justified on scientific or moral grounds. For that reason the book begins with a survey of the damage. Very few people can understand the intricacies of climate science although many of the basic principles can be explained to anyone who is seriously interested. Science is a part of the picture but not the most important part. Most people can live their lives knowing next to nothing about the science. But they are daily impacted by the cost of electricity and the fact that their children are being made into nervous wrecks by climate alarmism. This is a book for everyone because everyone is capable of understanding the damage that is being done when it is spelled out systematically and comprehensively.

After we survey the damage, we take a look at the supposed panacea of renewable energy, a resource that even some of the most ardent alarmists have recently come out against.

We then look at the German experience to show the failure of carbon mitigation even in the most capable hands. The Germans spent half a trillion Euros or more on global warming policies. This doubled the price of power but it seems that the only way they can reduce emissions is to go into recession!

The fourth chapter takes a counter-intuitive approach by explaining the massive benefits of the warming that we have enjoyed since the Little Ice Age and the significant greening of the planet in recent decades as the CO_2 level built up from the historically low level of 200 years ago.

In Chapter V we address the unproven theory that CO_2 is the principal driver of climate. Some readers might like to have the benefits of the warming chapter balanced by another to describe the downside of warming. However, we are not aware of any downside. We then have a look at what the main drivers of climate are and the almost imperceptible role of CO_2. These are concepts we should have learnt in school. For the older generation this will simply be jogging their memories. We suspect that for the younger people all this will be new information to them.

The organisation that collates and disseminates most of the global warming propaganda, the IPCC, is covered in the following chapter.

The computer models that have played a key role in the global warming movement are covered in Chapter VIII. Though computers are mere machines and are not subject to human fallibility, the people who program them and the data fed into them can certainly be manipulated into providing any desired result.

Chapter IX runs through a list of the common false alarms about warming.

The final three chapters describe how the "climate caper" is promoted, first by misleading language and images, second by proclaiming a "consensus" and then the big one – the money trail.

This book is not look like other books on Global Warming. This is intentional. Whilst there are many magnificent tomes on Global Warming from climate realists that we enjoy reading, most are not aimed at or written for the layman. One of the biggest issues for climate realists is that we prefer facts, observations and the scientific method which, sadly, can sometimes be hard-going for the layman.

This is also something cunningly exploited by alarmists who tend to talk in flowery, vague terms peppered with pseudo-scientific jargon, steeped in the language of Ivory Tower academia. The aim being to confuse or bore someone to death so that they will stop asking questions.

It is not like we have not seen this before in The Middle Ages. An elite class of experts, claiming to have exclusive knowledge, dispensed to the masses in a seemingly mystical language, unintelligible to the illiterate, Doomsday scenarios of fire and brimstone, the payment of indulgences and the casting out of "heretics".

This power over the minds of the masses was only broken when Martin Luther protested, Galileo was recalcitrant and William Tyndale translated the first English Bible. What followed was The Enlightenment and Age of Reason. As the magnificent physicist Richard Feynman said: "When we speak without jargon, it frees us from hiding behind knowledge we don't have."

Thus, we wanted this book to be unlike many other books on global warming. A book aimed at everyone from high school students to retirees, politicians with scant knowledge of the deception they are fed by lobbyists and those with an agenda and those that are just plain confused by everything they see and hear.

The Global Warming Clergy view the layman as being too ignorant and lowly. They feel that only they must be listened to and obeyed. This is not only elitist and exploitative but also self-serving and selfish. There is no monopoly on science or facts or knowledge, it does not belong to anyone or any institution, it simply is.

Our eternal gratitude goes to the team of contributors, editors, reviewers, fact-checkers and others who helped in any and every way to put this book together; Gilbert Mane, Charlotte Robinson, Diana Horska, The Climate Realists of Five Dock, encouragement from Will Happer and Peter Ridd and most of all to Rafe Champion who began this project and contributed much of the early work.

Chapter I
THE GOOD THE BAD AND THE UGLY
The Damage Caused by Climate Policies

"There is a great deal of ruin in a nation."

ADAM SMITH

All climate mitigation policies come with a cost, whether their advocates care to admit it or not. Even those who do count the cost usually do so in dollar terms. But there is another, potentially greater and longer lasting cost in the form of the collateral damage inflicted on humans and the environment. Some may find this more disturbing than the dollar cost. That human and environmental cost is essentially incalculable. How does one quantify the human cost of child slavery in the cobalt mines of the Congo, or the radioactive waste from the neodymium mines in Mongolia, or the land clearing in country Australia to install wind farms? If one *could* do a cold-hearted calculation of the savings on some or all of these climate interventions, the result would amount to some trillions of dollars every year worldwide. A vast sum of money that could be spent on facilities and services that might rectify some of this destruction that climate policies have produced and even enhance human health and welfare.

This chapter sketches the **human cost** in terms of lives lost or not being saved for want of reliable electric power that could be provided by hydrocarbon (fossil) fuels.

The **environmental impacts** include the loss of birds, bats and rainforests, the clearing of ecosystems to make way for solar panels and windmills and the looming problem of disposing of old panels and wind turbines with their highly toxic components.

The **social impacts** include the abundance of additional regulations; unjustified interference of the "nanny state" and encroachment and destruction of property rights. It also adversely influences public debate and journalism and diminishes the integrity of scientific knowledge, especially in schools.

The **economic cost** for Australia since renewable energy became a part of energy generation has been at least a doubling in the price of electricity. In some States it has tripled. There are calculations to show the cost of climate control policies in the United States and the worldwide cost of meeting the Paris agreements to reduce greenhouse gas emissions.

Human Cost

In the United States alone over 30% of grain production is taken out of the human food chain to produce ethanol to add to petrol, this amount is greater than the total grain produced by Russia annually[1]. This is called biofuel and it is supposed to reduce the emission of greenhouse gases from car exhausts.

A biofuel plant producing ethanol from the sugars found in grains such as corn and barley. [2]

There is no free market with biofuels; the European Union has mandated specific amounts of biofuel production, as has the United States, which also allocates subsidies, grants and tax credits. Diverting the corn from human consumption has a knock-on effect through the food chain, increasing the price[3] which has the most significant impact in the poorest parts of the developing world, where tens of millions of people are on the edge of starvation. A small difference in the price of food is a matter of life and death for them and there are estimates that in years of poor food harvests up to 200,000 people may die as a result of the biofuel program.

Mike Hulme is a leading international figure in the carbon mitigation movement so he cannot be accused of making up fake news to embarrass his colleagues. Hulme wrote in *Why We Disagree About Climate Change*:

> The UN's special rapporteur on the right to food, Jean Ziegler, condemned the growing of biofuels as 'a crime against humanity' because they diverted arable land to the production of crops which are then burned for fuel instead of sold for food...Rather than contributing to the theoretical food security of hundreds of millions of the world's poor in fifty years time, it has reduced the actual food security of tens of millions of the world's poor today.[4]

Nearly three billion people in the developing world cook and keep warm by burning wood and animal dung. This is classified as renewable energy called 'biomass' and makes up the vast majority of all renewable energy generated around the world. Biomass burned indoors generates toxic fumes causing lung diseases that kill millions of people every year, in the order of four million according to the World Health Organisation[5].

These lives will be saved when the people have access to electric power to eliminate indoor pollution. Electricity can be used to create the infrastructure required to deliver clean drinking water and public health programmes and vastly increase agricultural productivity and economic development. In most developing nations electric power could come from local resources of coal and gas but the international lending agencies were pressured by the Obama administration and the European Union to direct funding to renewable energy projects. This is part of the international Green movement to block investment in coal and gas projects not just in the West but also the developing world.

The fact that renewable energy cannot deliver enough power for household consumption, let alone industrial development was unwittingly demonstrated by a Greenpeace model village experiment in Dharnai in the North-East of India. They installed a solar powered micro-grid and when a government Minister arrived for the ceremony to celebrate green energy, he was greeted by rioting villagers demanding "real electricity, not fake electricity"[6]. The batteries only lasted for a few hours and the cost of the solar power was three times that of grid electricity.

How is it fair that a child in rural India must study in the evening under a kerosene lamp whilst Western children have desk lamps and bright LED bulbs? And an Indian Mother cooking dinner using animal dung rather than an electric oven or gas stove?

The late cartoonist Bill Leak ridicules "climate change activists who would send poor people solar panels rather than give them something they need - cheap power, aid and a hand up".[7]

Many millions of lives could be saved by using reliable electric power and thus releasing and diverting some of the billions which are spent on damaging and often ineffective carbon mitigation programmes in the affluent West. The money could then be used on conventional power projects in developing nations.

Some people think that there is no point in saving those lives because they will just starve to death anyway through over-population. This inhuman view does not take account of the phenomenon called the demographic transition that occurs when the populace reaches a certain level of development and start to limit the size of their families. This pattern has been repeated over-and-over again, most noticeably in China where the rising middle class usually limit their families to one or two children even though the One Child Policy is no longer in force[8].

Environmental Impact

Rainforests in developing countries are being cleared for jatropha, sugar cane and palm oil to convert into biofuel. Indonesia and Malaysia produce palm oil whilst jatropha and sugar cane are grown in sub-Saharan countries like Uganda and Mali as well as Pakistan, India and Central America.

The case for biofuel is very weak since it indirectly produces just as much CO_2 as regular petrol due to forest clearing, planting and harvesting the new biofuel crops and transporting the product.

Green energy policies too often produce wasteful and environmentally damaging outcomes. Many of these emanate from the illogical decision of the European Union to ignore CO_2 from burning biomass (mostly wood) in the score for national greenhouse gas emissions. The excuse is that the trees absorb CO_2 from the air when they are growing and the regrowth will take more CO_2 out of the air to balance the amount released by burning the wood pellets in the power station boilers. However, that process takes decades while forests in the U.S. and other countries are being clear-felled for wood pellets and transported thousands of miles (using fossil fuels) to be burned in British furnaces.

Chris Huhne served as the British Energy Secretary in the coalition government of the Conservatives and the Liberal Democrats from 2010-12. Huhne championed the use of wood pellets, notably in the gigantic DRAX power station that was converted from coal-burning to the burning of wood pellets, shipped from the United States. He advanced the use of household biomass furnaces in Britain with the *Renewable Heat Incentive* (RHI)[9] scheme. This allowed businesses and households to buy wood-burning heating systems and then receive payments for up to 20 years, depending on the amount of heat they produced. Some homeowners worked out that they could double the amount of heat they produced by using dry wood or other materials to feed back into the boilers and generate more payments. The scheme started in 2011 and it was open to applications until 2021. The payments to participants will run to 2041 and by that time the cost to taxpayers is expected to reach £23 billion. After Huhne left Parliament, he became the European chairman of Zilkha Biomass, a U.S. supplier of wood pellets.

The Irish "Cash for Ash"[10] scandal resembled the Renewable Heat Incentive scheme because it paid more than the cost of the fuel. And in Spain the generous subsidies for solar energy feeding into the grid incentivised some suppliers to shine diesel-powered lights on the panels after the sun set[11]. The solar powered torch joke is no longer amusing.

Those scandals shared similar features: the destruction of natural resources, waste of money and the compromising of business and citizen integrity when sections of the community "game the system". Another common feature is that these shortcomings are the results of schemes that are implemented without proper planning and due diligence. Scandals aside, misplaced priorities in one area often snowball into a widening circle of waste and damage in other areas. The cancer of climate alarmism infects policies at every level of government. The term "sustainability" is seen or heard in all manner of decision making, policy and public utterance. Entire government departments staffed by thousands of generously paid public servants are established to ensure that every public Agency at all levels of government is considering the alleged climate crisis in their plans and decisions.

Social Impact

Climate change policies have multiplied prodigiously in just a few decades, filtering into every facet of life. The Western countries are now Regulation Nations with overbearing Nanny States cajoling and pestering the people like a swarm of ravenous mosquitoes. To illustrate the escalation of rules and regulations driven by exaggerated climate and environmental scares, some 3000 pages of law and regulations were written to establish the renewable energy grid in South Australia. That became the model for all the Commonwealth and state regulations that followed. Globally, climate change laws since 1997 have increased from 60 to over 1260[12].

While on the one hand, serious environmental damage is being inflicted on the planet as described above. On the other, petty and inconsequential regulations are introduced that will make practically no difference at all to the environment while they cause inconvenience and irritation in our daily lives. Examples are the replacement of old style incandescent light globes and bans on plastic straws, balloons and plastic bags.

Climate alarmists emerged as a product of the radical shift from the sensible and environmentally sensitive conservation movement, into the radical Green movement who protest by gluing themselves to our roads today. The current vogue of climate-strike school protesters demonstrates the way the public discussion has been hijacked by unprincipled activists, who present only part of the argument. School children are being exposed to incomplete, inaccurate and misleading information on climate change and environmental issues in a very one-sided manner. There seems little attempt to present the issue as a scientific debate in which children are encouraged to make their own minds up using all the evidence. It is not hyperbolic to claim it is a dangerous and irresponsible way in which to impart knowledge and is comparable to the erroneous belief and claims of previous ages that the Sun orbits the Earth rather than vice versa.

Children are literally being frightened and stressed by Doomsday fearmongering presented to them in far too many aspects of their education and life. Reports of children feeling anxiety over the climate and requiring counselling, are distressingly frequent. This form of psychological manipulation by activists is doing tremendous damage to the psyches of young children who do not possess the intellectual rigour, training or evidence to question the authority of their elders and the media machine. The messaging to which children are exposed is reinforcing the false notion that the industrialised nations, primarily the Western nations, are inherently bad and evil. This helps to reinforce Cultural Marxist ideology and engender a sense of hatred for their country and culture.

Economic Cost

Due to the surge of investment in wind and solar power the published statistics on the cost and capacity of the renewable energy sector are often out of date by the time they are printed. Most of the actual cost is not "on the books" in Government budget papers because it is passed on to consumers in the form of higher prices for electricity.

Before we began spending on renewable energy, we had an abundant supply of cheap power. Now the injection of intermittent energy from the sun and the wind has driven the price of electricity even higher. This increased cost of power is also accompanied by reduced reliability of supply. Since 2010 no less than twelve coal power stations have closed in Australia, taking 5000MW (over 20% of total capacity) out of the system.

Hazelwood power station closed in 2017 after the Victorian government substantially increased the royalty charged for coal, the Energy Market Operator warned that we were dangerously short of spare capacity in the system. Blackouts in Victoria on 19 January 2019 proved the point. That has not happened again because the Operator has managed to summon up enough "Reserve Power" to scrape through tough periods. The Reserve Power is not really spare power, it actually comes from big power users who are required to cut back their consumption during the critical period. They have to be compensated by the Government. In other words taxpayers foot the bill.

The most visible cost for consumers is the price of power that has more than doubled since the introduction of the National Energy Market in the 1990s.

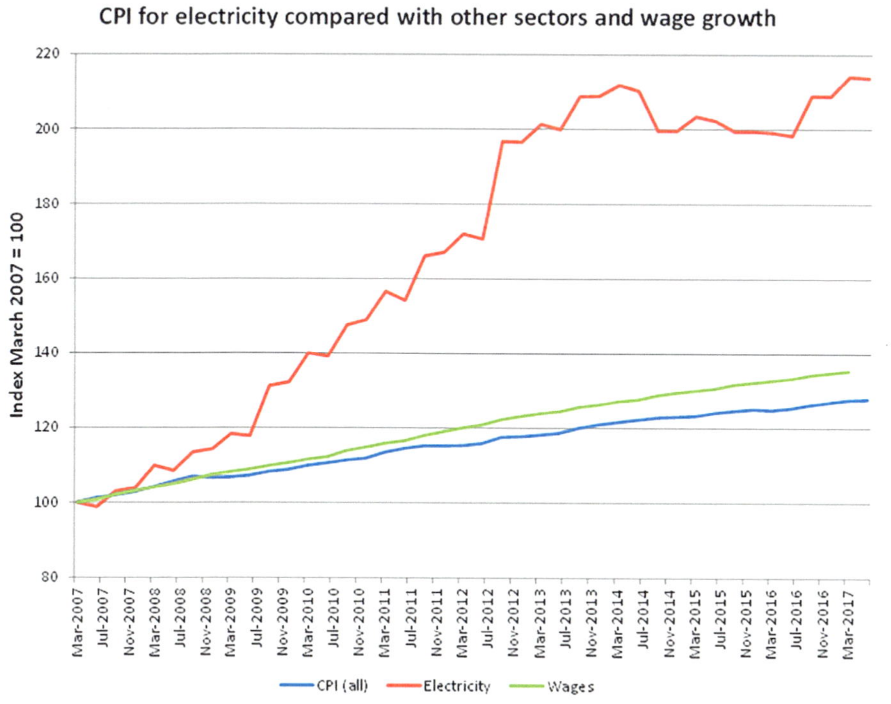

The steep part of the curve came when renewable energy entered the market at the turn of the 21st century. [13]

It is especially important to understand the impact of increasing power prices for low-income families and business enterprises that use cookers and refrigerators. For a low-income family, a significant increase in the cost of power means a move from the breadline to serious hardship. The great inventor Thomas Edison said in 1879 that "We will make Electric Light so cheap that only the wealthy can afford to burn candles", soon only the wealthy will be able to afford electricity and the poor will burn candles. As for small business, talk to your barista, local restaurant or pub and ask how they have been affected by power prices.

Economist Alan Moran uses the example of a 50% increase in the cost of power for a business where 20% of operating costs is electricity. The big hit comes on the bottom line of profits because it only needs a rise of 2% in costs to take as much as 13% out of the bottom line that keeps the workers employed and drives job creation in a vibrant economy. That hit translates into hundreds of thousands of jobs that will be lost if the renewable energy targets embraced by the major parties are achieved.

Electricity can be up to 60 per cent of costs in some enterprises and the first victims of higher prices are industries such as the aluminium smelters that add value to our mineral exports. Two smelters, Kurri Kurri and Point Henry, have already closed and in 2019 Rio Tinto warned it may close its three Australian Aluminium smelters because of power prices, leaving just the Alcoa Plant in Portland Victoria[14]. When aluminium ore ceases to be processed in Australia it will be shipped overseas, using oil, to countries like China and India who will use Australian coal, also shipped overseas using oil, that would have been used to smelt the aluminium in Australia. So Australia loses the jobs, the communities and the tax revenue while greenhouse gas emissions are increased.

In *The Hidden Cost of Climate Policies and Renewables*[15]Alan Moran reported that the cost of the subsidies for RE generators and the increased cost of power to consumers is:

- At least $13 billion per annum for households, near $1300 per household.
- 39% of household electricity bills, not the 6.5% quoted by the Government.
- 2.2 jobs lost for every "green" subsidised job that is created.

The picture gets worse for the economy as a whole. Independent analysis of the Labor and Coalition policies going into the 2019 General Election estimated the Labor Party plan to reduce emissions by 45% would cost $500,000,000,000 ($500 billion) to the year 2030 with the loss of 200,000 to 300,000 jobs as well as lower wage growth. The damage for the less ambitious Coalition plan was in the order of $90,000,000,000 ($90 billion) and 80,000 jobs. If The Greens had their way the Labor plan would have cost more than a $1,000,000,000,000 ($1 trillion) because The Greens would not allow the cost to be reduced by trading international carbon permits. Labor is still hostage to The Greens so nothing can be ruled out on the way to the next election.

In the United States author and President of the Copenhagen Consensus Centre Bjorn Lomborg did what the Australian Labor Party was not prepared to do. He carried out a cost-benefit analysis which he presented to the U.S. Senate Budget Committee in July 2014 in *The Costs of Inaction: The Economic and Budgetary Consequences of Climate Change*[16]. The bottom line was that the climate policies in place in the U.S. would cost much more than taking no action at all. The benefits of warming, discussed in the 'Some Like It Hot' chapter, have been calculated to exceed the costs of a 2°C warming. In other words, warming is a net benefit for most of this century as the prospect of a 2°C rise caused by CO_2 alone, as the climate change advocated allege, would not occur until well past the year 2100 according to their worst-case-scenario models. Lomborg believes in Anthropogenic Global Warming but owing to the fact he omits the Catastrophic 'C' in 'CAGW' he is treated like a pariah in the eyes of alarmists. So much so that staff and students twice successfully thwarted attempts to set up a cost/benefit program under his guidance at Australian Universities[17], the first of its kind anywhere in the world. This makes a mockery of the oft heard false claim that "the debate is over". The debate has never even been allowed to take place.

Of course, the costs only represent one side of the cost-benefit equation and benefits are supposed to come from reducing CO_2 emissions and limiting warming. However, Australia only contributes a fraction over 1% of total human emissions. The Chief Scientist of Australia advised a Parliamentary Inquiry that the proposals on the table to reduce our output would make no difference to the temperature of the planet. On top of that it is important to consider whether we can reduce emissions in the first place while we maintain our quality of life. What can we learn from other countries that have tried to effect a green energy transition and do we even know if green energy will work?

Chapter II
GONE WITH THE WIND
Why Unreliable Energy Cannot Work

"There is no more dangerous menace to civilization than a government of incompetent, corrupt, or vile men."

LUDWIG VON MISES

In this chapter we expose the alarming environmental impact of renewable energy in its various forms, contrary to claims that it is clean compared with coal and gas. We also explain that the intermittent nature of their input, plus the lack of large-scale storage means that it is physically impossible for them to replace conventional power to supply reliable and affordable energy.

The worldwide quest for "net zero" emissions is sustained by muddled thinking about renewable energy that epitomises the unscientific nature of climate alarmism. Energy cannot be "renewed", only transferred from one form to another but the renewable energy enthusiasts embrace the term "renewable" to convey the impression that the resources of solar power and wind do not run out, in the manner of the supposedly finite reserves of fossil fuels. That impression is false because solar power runs out at sunset every day, and the wind runs out on calm days and especially during prolonged wind droughts. Wind and solar can only provide intermittent power and they could well be called "unreliable energy."

The failure of unreliable energy in Germany is explained in Chapter III and there are many other examples including Denmark, California and South Australia, where the implementation of solar and wind have driven up costs and destabilized the grid.

There has been inadequate due diligence in government planning of green energy schemes – that means checking and re-checking the assumptions and the calculations. The costs, the resources required, and the power generated by solar and wind are well-known. Many individuals and even organisations as large as Google[1] have concluded that solar and wind cannot work as large-scale power generation sources. But still governments are ramping up their commitments to the target of net zero emissions.

In the beginning

Modern wind turbines were developed by NASA in the 1970s when the U.S Government was looking for alternative energy sources after the oil shocks of 1973 and 1979. The idea of using large wind turbines on towers to generate power was dreamt up decades earlier when the National Socialist (Nazi) Government was captivated by renewable energy technology and promoted it.

The prototype design in this picture[2] was never built owing to the lack of the necessary technology.

In the U.S., mostly in California, some attempts to build wind farms in the 1970s and 80s failed to meet expectations. They produced little energy and they were too expensive to maintain.

Solar was first used in a few toys and radios in the 1950s, after the invention of the silicon solar cell.

In the 1960s it was used on satellites but it was too expensive for large scale applications and it was confined to science-fiction:

When George Lucas wrote *Star Wars* in the 1970s he clad the Galactic Empire fighters in solar panels.

The evil Galactic Empire: Powered by solar.

Research spearheaded by Exxon (yes, "Big Oil") in the 1970s brought the price of solar down considerably, yet even then it was felt that solar would only have very limited applications[3]. Other oil companies invested in solar in the 1970s but after the end of the oil shocks and the relatively cold decade of the '70s they scaled back or closed their solar businesses.

Fifty years on there are now hundreds of thousands of wind turbines worldwide, alongside massive field solar plants and rapidly proliferating rooftop solar panels. This is a global frenzy, triggered by the climate alarm that provoked governments into providing subsidies, tax breaks and legislation to enforce the use of "clean" energy, regardless of the cost, the impact on the stability of the power supply and the environmental devastation that is only now being exposed.

The Environmental Impact

The myth that unreliable energy is environmentally "clean" is one of the Big Lies of our time. In the press, coal power stations are always described as "ageing", but wind turbines age a lot faster. A well-maintained coal station can run for 50 to 60 years and a nuclear plant can produce for 70 to 80 years but wind turbines rapidly lose efficiency after a decade and rarely last beyond 20 years. The relatively short lifetime of wind turbines appeals to the finance industry because the capital expenditure on wind power is required several times in the life of a single coal power station.

The most misleading claim by RE enthusiasts is that turning to wind and solar energy will save the planet from the impact of industrial development. A single wind turbine requires thousands of tons of concrete and steel as well as several rare earth elements that are mined in developing countries. The turbine itself requires magnetic elements like neodymium and dysprosium for the super powerful magnets that are also used, in tiny quantities, in cell phones and computers.

Nearly a ton of special purpose minerals is required for each turbine, mostly neodymium which is incredibly toxic to mine and refine. It is sourced exclusively from China and Mongolia and entire landscapes have been turned into toxic waste dumps that resemble dystopian futuristic nightmares. The Chinese Government maintains strict security around these areas and there are reports that workers and surrounding villagers are liable to suffer extremely debilitating diseases and increased cancer rates[4].

Mining these rare earth metals creates huge quantities of radioactive waste. Apparently nuclear power is bad for producing nuclear waste but not wind power. Considering actual generation and lifespan, wind turbines produce more radioactive waste than nuclear power plants that have well-established protocols for recycling and storage of waste.

Solar is not quite as environmentally problematic as wind, still there are considerable issues, many of which are not widely realised yet. Solar panels contain large quantities of lead, cadmium and other toxic chemicals. With lifespans of 15 to 25 years we will shortly face a real environmental problem when millions of panels are thrown away each year. At present there are no certified methods or regulations regarding solar panel disposal and it is likely the panels will end up in landfill where rainwater will carry carcinogens from the panels into the soil and groundwater. They may be dumped in the Third World where children will break them apart with crude hammers and tools, inhaling the toxic dust before the remnants are dumped into open landfill to poison their soils and waters.

The Footprint of Wind and Solar

Another negative impact of wind and solar facilities is that they use up vast areas of land. Solar farms require up to 75 times as much land area to produce the same amount of electricity as nuclear or coal power. Wind power demands up to 300 times the land area[5]. These wind and solar "farms" often come at the expense of real farms needed to grow food. Huge areas of land require clearing, with access roads to carry the massive pieces of machinery for wind installations. There will have to be more clearing for hundreds or even thousands of kilometers of high-tension transmission lines with associated substations to connect to the existing grid. The combination of road building, clearing and installation degrades and fragments wildlife habitats, impacts on water flow and promotes soil erosion.

The windmills have awesome dimensions that contrast with the picture postcard image seen from a distance. The latest models of wind turbines stand 220 metres high with blades up to 164 meters long. Blade assemblies can weigh 35 tons and the tower over 70 tons. The nacelle at the top is the size of a bus and weighs nearly 60 tons. The base can require 4,000 tons of reinforced concrete[6]. Contemplate the dollar cost and the CO_2 generated in the manufacture and transport of that amount of material multiplied many times over for a substantial wind farm! The footprint on the environment extends far beyond the sweep of the blades because a large turbine needs 30 to 40 clear hectares around it and the machines must be well spaced to minimize interference with each other.

One researcher calculated that to supply the whole of the UK with wind power would require the entire country of Wales to be covered with turbines[7]. In other words, you would have to destroy an entire country to power England and Scotland.

For some strange reason wind and solar farm projects are rarely subjected to the expensive and time-consuming delays and impact assessments that impede the progress of mines, dams and nuclear projects. We know that solar and wind have destructive impacts on wildlife due to the massive areas of land clearing and the millions of birds and bats that are killed each year by wind turbine blades and the concentrated reflections from solar farms[8].

The long-term impacts of wind turbines are not yet known and some environmental scientists have recently theorised that they may have a detrimental impact on the moisture content of the land. What are essentially thousands of giant open-air fans could be drying out the surrounding soils, usually arable farmland. Wind turbines are not best suited for a country prone to bushfires, droughts and which has inadequate infrastructure to get water to where it is needed most.

The 15-20-year lifespan of wind turbines is emerging as a huge issue because the blades cannot be recycled. There are over 50 tonnes of unrecyclable plastic in a 5MW turbine alone[9]. And our priority is to ban plastic drinking straws. In 2013 in the U.S alone there were 14,000 abandoned turbines[10] and it is estimated that tens of thousands of tonnes of wind turbine blades will be dumped into landfill over the next two decades. Currently, the owners of wind turbines do not have to pay anything upfront to cover the inevitable disposal and dismantling of turbines that have reached the end of their life. So taxpayers will have to pay up, again, for the extremely costly cleanup whilst also suffering from the effects of toxified soils and water when the blades are buried in landfill. [11]

The Environmental Costs of Renewable Energy

Part 1

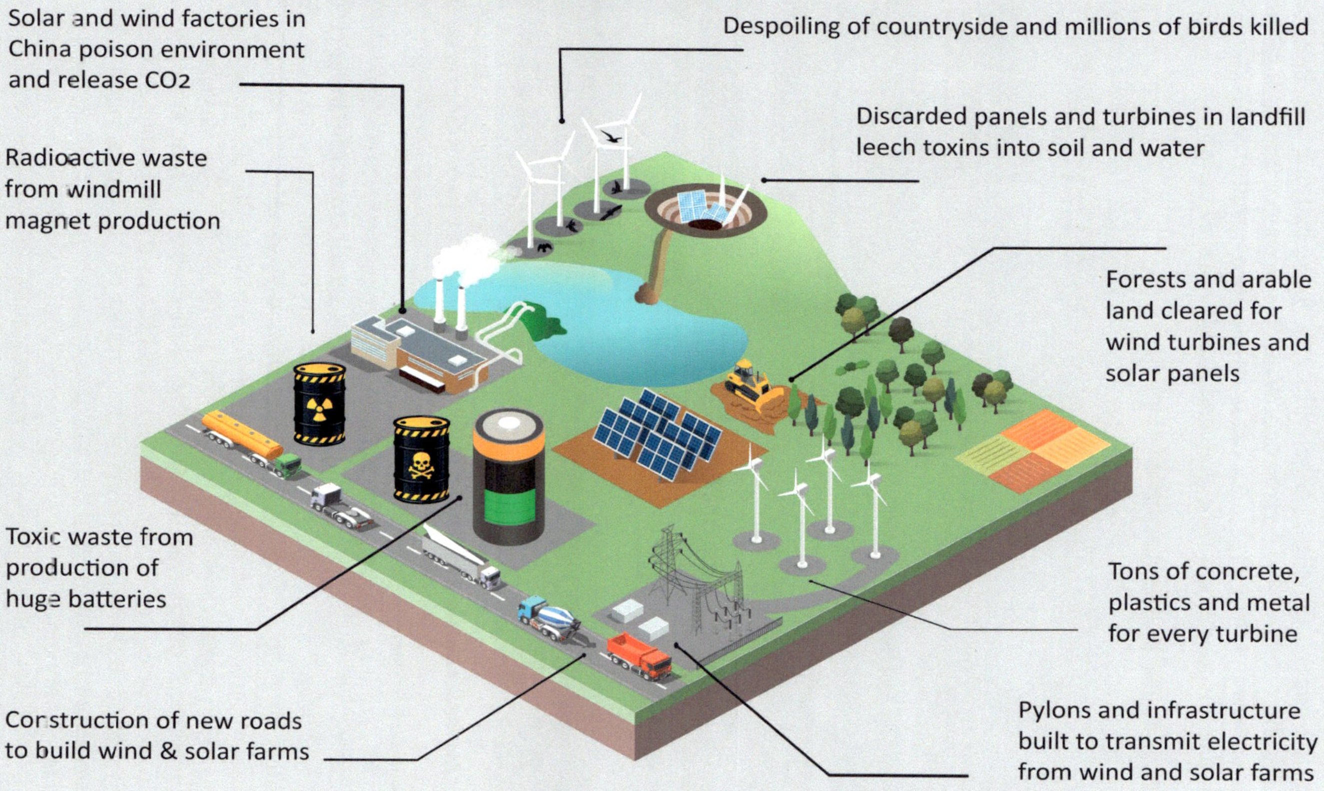

Bluff and Bluster

Solar panels do not stand up well to hailstones, and turbines can literally fly apart in high winds. According to the disturbing scenarios that we read about in the papers, climate change will cause more extreme weather conditions, including storms and hurricanes. With this prospect it is strange to see how much weather-vulnerable RE capacity has been installed, with swathes of land and rooftops covered with glass and forests of wind turbines.

Hurricane Maria hit Puerto Rico in September 2017 destroying hundreds of panels and wind turbines. This was shown in the 2020 Michael Moore documentary *Planet of the Humans*.

The Problem of Intermittent Input

Claims of the amount of power from windfarms are inflated by using "nameplate" values that are the maximum output capacity of the farm with ideal wind conditions. In fact, the average production in Australia is a little under 30% of the nameplate capacity because the wind supply is highly variable, from near zero up to the point where the turbines have to be "feathered" or turned off so they don't self-destruct. These low periods can last from a few hours to days, delivering as little as 1% of nameplate capacity. That is next to nothing, regardless of the number of windmills. Wind is like the fuel in the petrol tank in your car; it will not run when the tank is empty, even if the tank is very large.

Windless nights present an insurmountable problem for wind and solar power. When the sun sets there is no more input until next day and on windless nights there is no wind power either. The critical question is: How often and how long do the wind turbines stand idle, or near idle?

The answer is readily available in the records of the Australian Energy Market Operator (AEMO.) These show that the wind power supply frequently drops below 10% of the nameplate or potential capacity of all the wind farms in South-eastern Australia. This is likely to happen in any month of the year and especially in the autumn and winter months. For example, in June 2020 the supply fell below 10% of potential capacity on twelve occasions. The longest period was 33 hours and others were 18, 16, 12 and 9 hours. This is the area shaded in red. The green shaded area shows that capacity very rarely gets above 50%.

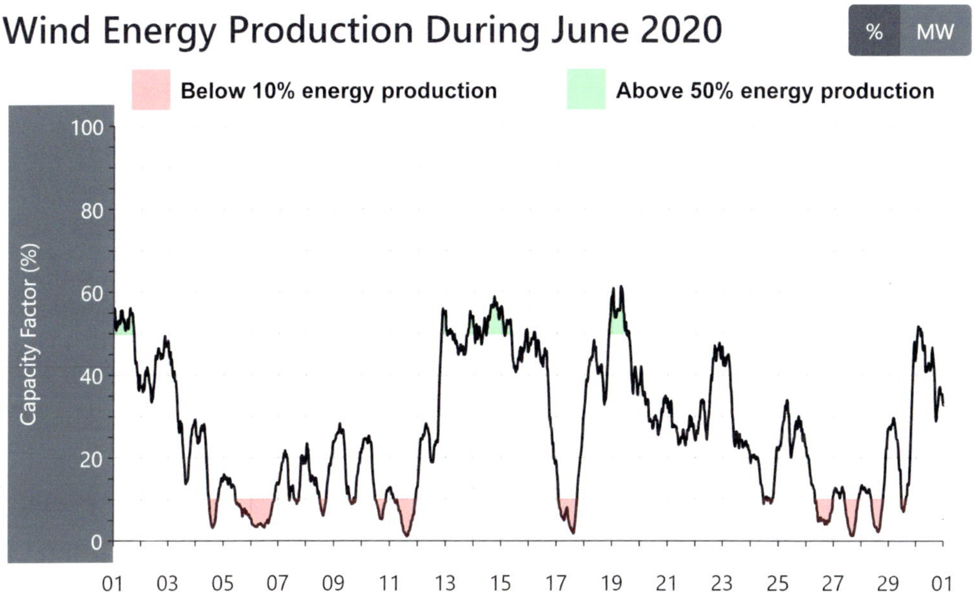

This means that RE alone cannot provide a reliable supply of power around the clock every day of the year. In the absence of conventional power, the lights will go out every time there is a prolonged wind drought, that is, several times a year. No amount of additional capacity will help because the low points of the supply kill the grid, in the way that a low point or a gap in a flood levee lets the water in, regardless of the high points or the average height of the defensive wall.

Of course, the planners realise that there are low points in the supply and they offer three solutions to maintain the supply of power: more transmission lines, batteries and pumped hydro.

Transmission lines

Transmission lines are expected to carry wind and solar power from places where there is a surplus to places where there is a deficit. This will not help when there is a wind drought across the whole of SE Australia and this is the case whenever high-pressure weather systems linger on their journey from west to east across the continent. The chart above shows that during the low periods there will be no surplus wind power to transmit.

Big batteries are not big enough

Batteries work for small devices like mobile phones and laptops and you can spend tens of thousands of dollars for a household battery. However, at the grid scale they run into the issues of environmental impact, lifespan, cost, safety and capacity.

The main components of batteries are lithium and cobalt, a rare earth metal mined predominantly in The Congo. This mining is done in very harsh conditions, at subsistence level wages and using child slave labour, children as young as four have been found to be working in the mines[12].

Batteries in mobile phones and other devices lose their capacity after a few years. When you buy a new phone, a single charge can last for days but after about two years you will be charging it nearly every day. This issue also exists with large batteries whether they be in electric vehicles, bolted to the side of your home or at a battery farm in South Australia. Larger batteries also tend to be more unstable and when they catch fire can be very difficult to extinguish. Having thousands of these things spread out across fire-prone areas is a huge risk.

There is a lot of publicity about "big" batteries that are being installed, following the lead of the much-publicised Elon Musk battery attached to the Hornsdale wind farm in South Australia. This was supposed to keep the lights on in South Australia but it will not overcome the problem of wind droughts because its storage capacity is vanishingly small compared with the demand of the grid. Its capacity is 194 Megawatt hours (MWh) compared with the demand from the grid that is near 1,500MWh every hour. The battery will last about six minutes if the State loses power again like it did in September 2016.

The battery occupies a hectare, slightly larger than a sports field, and it cost over $200 million – that is over a million dollars per MWh of storage. Its real function is not bulk storage but providing a short, sharp injection of power to balance a sudden fall in production from the wind farm that can destabilize the grid.

The false promise of pumped hydro

Pumped hydro power is generated when water from an upper reservoir flows through generators to a lower reservoir. Power from wind and solar farms is used to pump water from the lower reservoir back to the top, consuming some 30 to 40% of the original power in the process. The massive and environmentally damaging Snowy 2.0 project will replace almost one large coal-fired power station. The capital cost is expected to be $10 billion or more plus the cost of 8,000 to 10,000MW of wind and solar capacity to provide the input. Eight or nine more projects of the same size would be required to replace all current coal power but the cost and environmental impact will almost certainly block further developments after Snowy 2.0.

RE is surplus to requirements

The vagaries of the RE supply did not matter when we had enough conventional power, mostly from coal, to keep the grid alive. However, with twelve coal power stations closed since 2010, conventional generation capacity has decreased 20%. After the Hazelwood power station in Victoria closed in

2017 there has not been enough spare capacity to cope with exceptionally large demands during heatwaves and colder Winters. In motoring language, we are travelling on the rims of the wheels with no spare tyre in the boot. In January 2019 there were blackouts in Victoria during a heatwave. In the middle of May 2021 there were repeated shutdowns at the Tomago smelter during a wind drought when a single generator (one of four) went down at Liddell power station in New South Wales. In 2023 more generators at Liddell are scheduled to close and that will create an unprecedented situation in the power industry because regular RE input will be necessary to keep the lights on.

Under these circumstances it is clearly suicidal to close down conventional power, which essentially means coal power, until there is some replacement that can run for several days rather than a few minutes or, better still, be available 24 hours a day as required. No amount of installed capacity of wind and solar will help until there is feasible and affordable grid-scale storage. Until then we will need 100% backup from conventional power, and all our wind and solar capacity will be superfluous to requirements. This means that it was a mistake to connect RE to the grid and it should have been left to businesses and individuals to use wind and solar power off-grid if it is essential or cost-effective for them.

RE can displace coal power but not replace it

As long as 100% backup is required, solar and wind facilities can only be seen as expensive, virtue signalling ornaments attached to the grid. Tragically they are not just ornamental. Supported by subsidies and legal requirements that mandate the use of RE, wind and solar power is displacing coal by bidding lower in the wholesale market. The coal providers are being driven towards financial collapse, several have already gone and when the next coal plant closes RE will have to step up and replace coal instead of just displacing it. Given the phenomenon of wind droughts and lack of grid-scale storage, that is not possible.

Alternative sources of "clean" energy

Many attempts to source RE energy have failed spectacularly, ranging from "hot rocks" to various wave-powered generators that are now rusting in the sea. The 2020 Michael Moore documentary *Planet of the Humans* revealed how "biomass", the major form of RE, means clear felling forests in North and South America to ship wood pellets overseas to feed into converted coal-fired power plants in Europe. This vandalistic and wasteful practice exploits the EU regulations that don't count CO_2 emissions from burning biomass in the national emission score.

The most obvious sources of "clean" energy are nuclear and hydro but unfounded concerns about nuclear power have so far ruled out this option in Australia. Hydro is dismissed as being too environmentally destructive and the IMF has refused to fund hydroelectric dams in Africa leaving China to step in and fund these projects.

The Environmental Costs of Renewable Energy

Part 2

Nacelle with neodymium magnet. Production of neodymium creates over 100 times more nuclear waste than a nuclear power plant

A standard turbine contains over 50 tonnes of unrecyclable plastic in the blades

Uses over 800 tons of steel produced with coking coal and iron ore from mining

Short lifespan of 20 years after which they are often left to rust or buried in landfill

4,000 ton concrete base

Made of toxic lead, cadmium, hexafluoroethane, polyvinyl fluoride and silicon tetrachloride.

Need replacing every 15-25 years, cannot be recylced.

Create 300 times more toxic waste per unit of energy than nuclear power plants.

Release nitrogen trifluoride, a chemical compound 17,000 times more potent than carbon dioxide.

Large scale batteries contain rare earth minerals: cobalt, lithium, nickel

Minerals like cobalt are mined in the Third World using slave labour

Processing of metals produces a lot of toxic waste

Short life span of around 10 years largely unrecyclable

Nuclear power is a proven technology that was delivering large scale, affordable energy long before wind and solar came along. France is over 70% nuclear and household prices are about half those in Germany. Even in the countries where nuclear power is legal governments have invested next to nothing in research into thorium reactors or even the most desirable form of energy production, nuclear fusion. This is partly under the influence of Greens who detest nuclear power (and most other forms of conventional power as well). Here The Greens are supported by the gas lobby. Gas can thrive alongside the intermittent providers because it has to be called up when the wind and sun are not delivering. Gas competes with the reliable providers, especially coal and nuclear power and so they add weight to the anti-coal and anti-nuclear lobbies.

Coal seam gas (CSG) is another option and the U.S. is practically the only country to achieve any significant reduction in greenhouse gas emissions thanks to the use of innovative "fracking" to release gas from massive reserves locked in shale rock formations.

CSG pre-dates wind and solar, and in Australia the first fracking well was sunk in the Sydney suburb of Balmain in the early 1900s. Recent developments have been blocked on the basis of spurious claims of dangers despite ample evidence that it is quite safe.

In the U.S. the CSG boom delivered an estimated three million jobs by 2020. Meanwhile, the demand for rare earth metals for wind turbines and electric vehicles will see an increase in child slave labour in Third World countries. In the age of Fair-Trade Coffee, campaigns against Blood Diamonds and boycotts of companies using cheap labour in Asia to make running shoes, you would think there would be more news reports about the fate of the Congolese child slave labourers.

Coal seam gas workers in the U.S earn over $700 per day. Children mining cobalt in The Congo earn about 10 cents a day.[14]

The emerging crisis of grid stability

Solar and wind deliver intermittent and sometimes rapidly fluctuating inputs to the grid unlike the conventional providers that can run steadily, ramping up and down to match the predictable daily cycle of demand. The voltage in the electricity supply needs to be stabilised at 50 Hz because variations can result in serious damage to anything plugged into the power socket. There is very little room for crror either side of this figure and the grid can collapse in just a few hundredths of a second as occurred in South Australia in 2016 when wind capacity dropped suddenly, the frequency dropped to 47 Hz and the interconnector to Victoria was shut down, blacking out an entire State[15]. Conventional power does not have this issue as the turbines spin at a certain speed (frequency) and there are inbuilt mechanisms to instantly correct deviations.

Robbing Peter to provide power for Paul

Faced with the threat of serious blackout during periods of peak demand the Australian Government came up with the Reliability and Emergency Reserve Trader (RERT) scheme. This suggests that there is a reserve supply of power, like a reserve tank of petrol in a car but it really means that selected high consumers of power are required to reduce their consumption during critical periods. This is not a transparent process and the compensation that they are paid is added to the cost of power for all customers. Forcing industry to shut down to maintain supply elsewhere on one particular occasion cost taxpayers upwards of $24 million[16].

The RERT is supposed to be a voluntary scheme but on January 24th, 2019, the Australian Energy Market Operator forced the Alcoa aluminium smelter in Victoria to shut down without warning and without compensation[17]. The peak demand on that day pushed the market price for electricity to $14,500 MWh (the average price for 2019 was $90 MWh) but there was still not enough power and rolling blackouts in Victoria impacted over 200,000 people[18].

Industries were forced to shut down without compensation, costing them millions, the power went out anyway and the price paid for just **one day** of intermittent electricity was almost $1,000,000,000 ($1 billion). It's as if the Australian grid is run by Dr. Evil.

There is a serious lack of transparency in the system and the public cannot get a clear picture of the costs. Much of the increased cost of power caused by government policies is "off the books" and one example is the 2016 *Safeguard Mechanism* legislation. After the Australian Government dropped the highly unpopular Carbon Tax, they legislated for the power companies to effectively collect the revamped Carbon Tax on their behalf and hiding the cost to consumers in their power bills.

Social Justice and Rooftop Solar

It is surprising that the people who are most vocal about "social justice" issues have yet to agitate about one of the most glaring examples of redistribution from the poor to the rich. This is the rooftop solar subsidy system and the feed-in tariffs. Taxpayers fund the subsidies for people who can afford solar panels and batteries and own a house on which to install them. Then other electricity consumers fund the feed-in tariffs which is the money customers receive for any unused electricity that their solar power system sends back to the power grid. That is a win/win for some and lose/lose for others who can't access the subsidies and tariffs, either because they can't afford the upfront payment, don't own a home or their property is unsuitable (apartment owners). This wasteful and socially regressive scheme has prompted a frenzy of rooftop solar installations that has put Australia ahead of the world in the rate of installation but has also generated serious problems of grid stability that will get worse.

Conclusion

The development of the Australian power system in recent years offers a very sorry story. The price for consumers fell significantly from the 1950s to the 1980s and remained stable until the turn of the millennium when it "hockey sticked", increasing rapidly in a short period, mostly due to the increasing supply of RE. While the grid becomes less reliable the long-promised lower retail power prices are still nowhere to be seen, but the states and the Commonwealth are committed to more of the same policies that have made electricity more expensive and blackouts more frequent. History will not be kind to those who have brought about this situation.

Chapter III
DOWNFALL
The German Trifecta of Failure

"Some ideas are so wrong, that only a very intelligent person could believe them."

GEORGE ORWELL

Energiewende is German for "Energy transition" which describes the transition from coal to green energy. No country has more scientific and engineering capacity to enable this transformation better than Germany and there is a lot to learn from their experience. This energy transition had three main objectives. They were to attain an energy supply that was: stable, cheap and climate-friendly. These three objectives were referred to as the three sides of the "energy policy target triangle". The truth is that they failed on all three of these aspects of their plan.

The 6th *Energiewende Monitoring Report*[1] appeared in 2018, after a delay of two years, with a clear message of failure on supply security, affordability and emission reduction. The damning report was replete with criticisms about "unprecedented" wastage of resources, "missed targets" and "loss of confidence"[2]. As an official Government report, the negative findings cannot be dismissively attributed to critics of the scheme.

In terms of stability the failure of the green transition is comprehensive. The grid staggers on the brink of collapse from any combination of unfavourable conditions; minimal wind at a time of maximum demand plus the failure of some coal-fired capacity. That problem will only get worse because Germany's nine remaining nuclear power stations are scheduled to close by the end of 2022[3]. A hidden cost of the uncertainty of supply is the hundreds of millions of Euros that firms in power-sensitive industries have spent on diesel generators for backup in case the electricity supply fails.

On affordability, Germany and Denmark shared the World Cup for the highest energy prices for some time before South Australia overtook them both in 2018[4].

Clearly the more renewables that are installed, the higher the cost of electricity. This contradicts everything that the media supporters and backers of renewable energy repeatedly claim. Prime Ministers have made hopelessly inaccurate claims that renewables would cost people "$1 Per year"[5] and regularly state that "prices will begin to fall in a few years". Perhaps prices will fall one day but something that previously cost $500 and then increases to $1000 and later falls to $900 has not "become cheaper". The 340,000 German households who had their electricity shut off in 2018 because they were unable to pay their increasingly expensive bills are still awaiting this nirvana of cheap renewable energy[6].

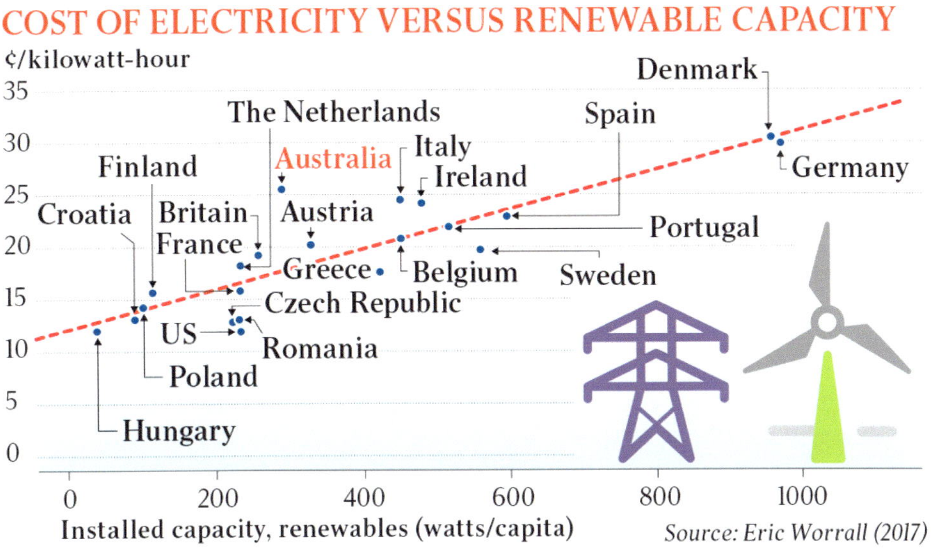

This chart demonstrates that as more renewable capacity is installed, electricity prices increase. Australia is amongst the highest prices in the world for renewable energy per kwh.[7]

The aim of the *Energiewende*, its entire *raison d'etre*, is to reduce emissions of CO_2. It has completely failed to even come close to meeting the target. The ardent green energy lobbyists could not hide this fact in the chart produced in the 2018 edition of *A Reporters Guide to the Energiewende*.

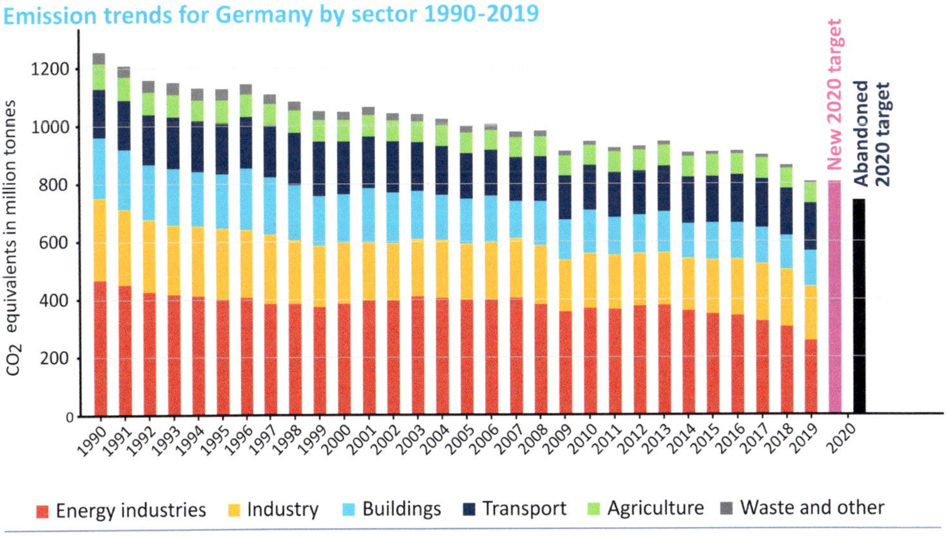

After the modernization of industry in East Germany and the recession in 2008, progress in reducing emissions has been very small and far short of the target set in 2011.[8]

All pain and no gain.

The decline in CO_2 emissions from 1990 to the early 2000s came from the replacement of old and inefficient industrial plant in communist East Germany after the fall of the Berlin Wall and re-unification. The Global Financial Crisis in 2008-09 depressed industrial activity and produced further "progress" in reducing emissions. From that point the chart is flat with no reduction manifest for the best part of the decade. Ironically, having taken a step forward three decades ago German industry is now forced to use an electricity grid that more closely resembles a Trabant than a BMW.

The Origins of the *Energiewende*

The story of the *Energiewende* begins in the 1980s with the formation of The German Greens, the first official Green Political Party in the World upon which all other Green parties are modelled. Amongst those instrumental in the establishment of the Party was Günther Schwab, founder of the *World Union for Protection of Life*. Schwab was the former Nazi, eugenicist and author of the 1958 novel *Der Tanz mit dem Teufel* (The Dance with the Devil).

The German Greens started to talk about a transition to Green Energy without CO_2 emissions and in the 1990s their political base grew to the point where they joined the coalition government with the Social Democratic Party of Germany. They pushed their Green energy policies and in 2000 Germany passed a comprehensive Renewable Energy Act. The German Greens had pushed for an end to reliance on nuclear since their inception and seized on the flooding of the Japanese Fukushima nuclear reactor in 2011 to demand an end to nuclear power in Germany by 2022, even though Germany is not susceptible to earthquakes and tsunamis. Several plants closed almost immediately and a timetable was set for the closure of the others.

In 2007 Angela Merkel, the German Chancellor, was the driving force in the European Union due to the dominant position of Germany and she pushed ambitious green targets for the Union. The recurring theme was '20'; the target date was 2020, the target for renewable energy was 20% and the target for greenhouse gas reduction was 20%. Germany led the way and other places around the world followed enthusiastically, notably Spain, Denmark, Britain, South Australia and Puerto Rico. In 2011, under pressure from the Greens to be more ambitious, the Germans set new targets for greenhouse gas reduction; 40% by 2020, 55% by 2030 and up to 95% by 2050.

As power began slipping from Chancellor Merkel's hands in 2017 a new coalition formed and the powerful Trade Union movement was unhappy with the way the scheme was working out for their membership. In the early days the Trade Unions bought into the dream of green jobs but as power prices increased the Chinese undercut the price of solar panels and 130,000 solar panel jobs went to China. The same process was happening with the wind turbine industry. The cost of green jobs turned out to be enormous. Generally, each "green" job costs two or three ordinary jobs. In the case of the work on the fleet of offshore wind platforms the costs were even more expensive owing to the highly specialized skills, training and stipends for hazardous work conditions.

Looking again at the chart to track the reduction in CO_2 emissions it is apparent that the effort since the new targets were set in 2009 has been futile and the huge sums of taxpayers money spent have achieved nothing. The future of the *Energiewende* is now in limbo. It is hard to see how Germany can discontinue using coal power and close their remaining nuclear power stations without shedding power-intensive industries and damaging their economy.

What can others learn from the German Trifecta of Failure? Australia should be taking very serious notice since she does not have many of the advantages of Germany such as nuclear power stations, access to the power grids of their neighbours like France, who have nuclear power, and others with large amounts of hydropower like Norway, Sweden, Italy and Austria.

The State of New York State

The German experience was replicated in New York State where Audrey Zibelman Chaired the New York Public Service Commission that presided over the regulation and safety of the electricity, gas, telephone, cable and water system and set consumer rates for electricity. The Empire States *Reform the Energy Vision* plan would, in theory, draw all their power from natural gas, local nuclear facilities and local hydro, and use out of state sources for the remainder. Their aim is to increase to 50% renewables by 2030, close one of its four nuclear facilities and the two-remaining coal-fired power stations.

Zibelman claimed that moving toward a system that reduced carbon emissions did not necessarily mean higher costs. "It actually means lower prices if we do it right," as reported in a story on the New York electricity system[9].

In fact, just the announcement of the impending shuttering of the coal stations sent the electricity market slightly crazy[10] with the price of power surging more than 30 percent. The New York Times reported; "New York is taking lessons from California, Germany and other clean energy pioneers."[11] The Empire State can be added to the growing list of regions that seemingly want to bring about the Age of Endarkenment.

In the private sector Ms Zibelman was a founder, past President and CEO of Viridity Energy, not an energy provider but a tech company that develops software necessary to manage grids destabilised by the inputs of wind and solar. Owing to the unreliability of wind and solar, where output can jump wildly in seconds, the technological infrastructure required to manage supply and demand around renewable grids is complex and expensive and now with batteries being deployed as backup the complexity has vastly increased.

Zibelman was tipped to lead the United States Department of Energy if Hillary Clinton became President in 2016[12]. Instead Zibelman came to Australia in 2017 to head up the Energy Market Operator (AEMO) to guide our transition to green energy. Unfortunately, Ms Zibelman appears to have implemented the New York plan in Australia despite the radical differences in the Australian grid infrastructure: no nuclear power, nothing like 20% hydro and no connections to any external provider.

After arriving in Australia, at a conference held at the University of NSW, she repeated the mantra "you can have it all" – cheap power, a reliable grid and greatly reduced CO_2 emissions – provided that we do it right[13].

Naturally, when none of this occurs Zibelman will simply claim that "it wasn't done right".

The Australian Quadfecta: Planning to Fail

Not to be outdone, Australia has gone one better than Germany by building a quadfecta of failure. AEMO is just one of four organisations that run the Australian grid.

The Australian Energy Market Operator (AEMO) has the operational responsibility for the electricity system and also models scenarios to implement the green energy transition. Hence the CEO is widely regarded as the Energy Tzar of Australia. The Australian Energy Regulator (AER) oversees the operation of the wholesale and retail energy markets under the national energy rules. The Australian Energy Market Commission (AEMC) makes the rules and provides advice on energy market developments. The Energy Consumers Association (ECA) is the national voice for residential and small business energy consumers. The Energy Security Board (ESB) represents the other agencies to provide strategic advice to Governments on energy policy. Under normal circumstances this agency would probably not be required but the emerging crisis in the system has recently placed it in the spotlight.

Along with Ms Zibelman at AEMO from 2017 to 2021, the ECA Chair is Louise Sylvan with Rosemary Sinclair as CEO to 2021. The AER is Chaired by Clare Savage and the CEO is Michelle Groves. The AEMC is Chaired by John Pierce and the CEO is Anne Pearson. Three of the leading energy organisations in the country are run by people who have no qualifications relevant to the electricity industry. Zibelman has degrees in arts, law and business, Sylvan graduated in Public Administration, Savage is a career public servant with qualifications in commerce, history and politics, Groves is a lawyer, Sinclair and Pearson have degrees in arts, law and business and Pierce graduated in economics. The ESB is Chaired by Kerry Shott who has arts degrees, no experience in energy and was previously an academic, public servant and investment banker. Once upon a time, our energy industry leaders were actually qualified and experienced in engineering and physics.

It gets worse. Of the nearly sixty people who make up the Boards and Senior Management of The Australian Quadfecta, it seems there are only nine people with qualifications in electrical engineering or with scientific degrees. The rest are a mixture of lawyers, career public servants and former politicians with a range of qualifications from social work to horticulture and indigenous studies.

Evidently, the people running the Australian electricity sector are highly educated but there is a distinct lack of people with physics or engineering qualifications. Is it any wonder the situation is so dire when our electricity grids are no longer managed by engineers but by Green lawyers and bureaucrats who think an ohm is something chanted at a meditation class?

Chapter IV
SOME LIKE IT HOT
The Benefits of Warming

"Warmth is the vital element for the growing plant and for the soul of the child."

CARL JUNG

Every educated person knows that the globe has experienced cycles of warming and cooling for countless millions of years. When we look at the past these cycles of warming and cooling paint a noticeably clear picture; the Earth and its inhabitants flourish during warming and suffer during cooling. There is a reason life is abundant around the equator where it is warmer than the poles. In the tropical rain forests vegetation is luxurious and grows quickly to occupy any clearing or abandoned building[1].

What role CO_2 plays in warming is still very much up for debate, despite what we have been told. However, the notion that all warming is bad and will lead to terrible consequences is simply not supported by written history or the archaeological record or indeed common sense. It is therefore crucial to look at the science and historical evidence to determine whether a warmer Earth would be a net benefit or a calamity to humanity and the vast diversity of fauna and flora.

Let us begin by looking at the diagram which we will call The Global Thermostat[2].

The Global Thermostat

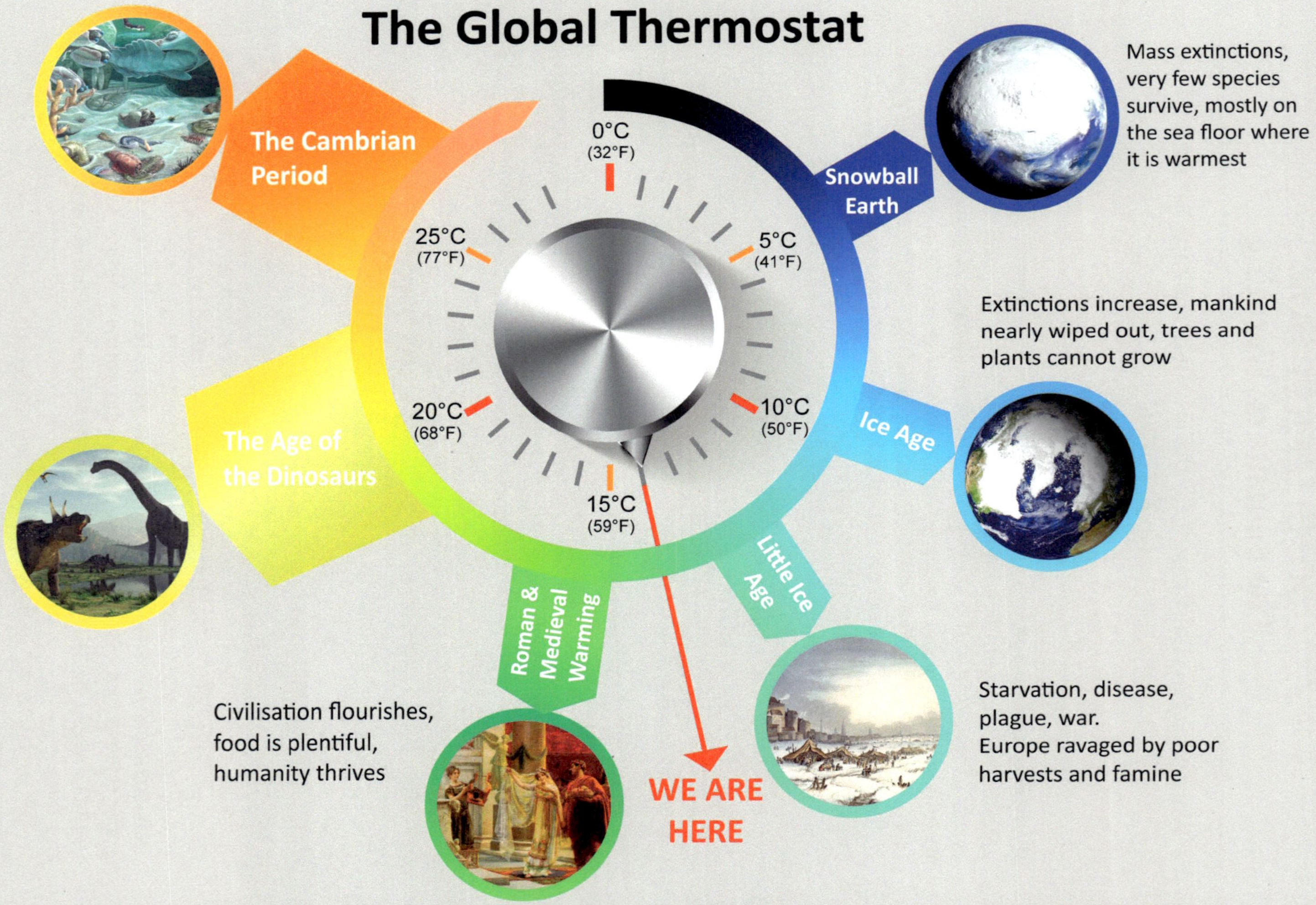

The dial is currently at 14°C which is the approximate global average temperature of the Earth at present. The idea of a global average temperature is regarded by some scientists as irrelevant and meaningless. There is a lot of truth in this. The Little Ice Age, for example, hit the Northern hemisphere much harder than the Southern hemisphere where temperatures varied less. The discussion around global average temperature is too long and complex for this book even though it is heavily over emphasized by alarmists. The actual topic we will cover in this chapter is simply whether warming is good or bad.

The Global Thermostat covers some of the major events in the history of life on Earth, called The *Pharenezoic Eon*, which stretches back some 550 million (half a billion) years. This was when single-celled bacteria evolved into multiple diverse organisms, to which all animal phyla that exist today can trace their origins back. This so-called Cambrian explosion happened when both temperature and CO_2 were considerably higher than today. The oceans, where life really flourished, were also more than 5°C warmer[3] yet it is the view of some people that a 1°C rise will destroy coral reefs and acidify the oceans. Those who claim that a temperature rise of one or two degrees will cause mass extinctions, clearly have a lack of understanding or knowledge of the historical record of planet Earth. There have of course been major events which have resulted in the extinctions of life throughout the past half billion years; five to be exact. There is still much debate about the reasons for these extinctions. Massive volcanic activity, asteroids, acid rain and Ice Ages have all been advanced as likely causes, either on their own or occurring concurrently or consecutively.

The world of trilobites and triceratops was obviously different to that of modern man though when we look back over the past 20,000 years or so we can see that people and nature both flourish in warmer temperatures. A much better understanding of the climate is acquired the closer we get to the present day. The records preserved in ice cores, the ground, fossilized plant and animal remains, genetic footprints and eventually written records, give us quite a clear picture. The peaks of the last Ice Age about 70,000 years ago nearly drove humans to extinction and it was only when the Earth warmed up and the glaciers retreated that humans were able to flourish.

One extraordinary event around 13,000 years ago highlights how rapidly the Earth's climate can change naturally. About 17,500 years ago the earth emerged from the coldest period of the current Ice Age, the sixth one, which began around 2.6 million years ago. This great thaw was interrupted by The Younger Dryas event when temperatures rapidly cooled again for about 1,300 years causing sea level to drop dramatically, the extinction of most megafauna and a reduction in human population. The Younger Dryas ended even more rapidly than it began, warming as much as 10°C in one decade or about 1°C every year in certain areas. There is still debate over what ended The Younger Dryas event, so far nobody has blamed SUV's or flatulent mammoths. Considering the debate over this event and other periods of Earth's history it is clear we do not fully understand the Earth's climate. Today, some claim that the 0.7°C warming over the last 100+ years is unprecedented and has never been seen before. This is completely erroneous, there have been multiple periods of rapid warming and cooling throughout Earth's history.

So how do scientists claim "the science is settled" over man-made climate change? How are they certain that current changes in climate are not driven by natural processes? Why are they so certain the changes are solely caused by man-made CO_2?

The warmer climate that began around 11,000 years ago was also the dawn of human civilisation, first in Mesopotamia followed by ancient India and China. Mankind planted and developed crops, devised irrigation systems, domesticated their animals and later built towns and cities as the climate became warmer. Anthropologists have speculated that the previous millennia of cold conditions were the impetus for these developments; mankind wanted to store food and have shelter should the cold return.

The Cool Holocene

This current period known as the Holocene epoch is not a particularly warm period. We are, after all, still in an Ice Age. This may come as a surprise to many people who tend to think of Ice Ages as woolly mammoths stomping across frozen tundra. It is quite simple to tell if you are in an Ice Age; if there is an ice sheet at the North Pole and the South Pole is covered in snow and ice then you are in an Ice Age; there have been no polar ice caps for the vast majority of Earth's history. The part of the Ice Age we are in is the warmer part known as an interglacial. In other words, we are in between the periods of glaciation when masses of ice and glaciers cover the Earth. There is plenty of time to prepare before the ice returns, roughly 23,000 years[4].

The Ice Ages were challenging times for living things and we have only recently emerged from the Little Ice Age that extended from about 1300 to the peak of The Industrial Revolution in the nineteenth century. This was when the Earth almost dropped back into a full Ice Age, a time of persistently bitter cold weather, that rapidly and dramatically replaced the Medieval Warm Period (MWP). The warm period from 900 to 1300 roughly coincided with the Middle Ages in Europe and the temperatures were warmer in most areas than the present time. The Norse colonised Iceland and Greenland and the favourable conditions at the time encouraged an increase in farming and resulted in a period of prosperity reflected in the building of the great cathedrals of Europe. Everything changed in the Little Ice Age that started around 1290 continuing until the 1800's. There was a slightly warmer period in the 1500's and then the climate deteriorated dramatically with the coldest period between 1645 and 1715. At the most bitter phase of the Little Ice Age the average winter temperatures in Europe and North America were as much as 2°C lower than at present. It was common for the River Thames to freeze over in the Winter and "Frost Fairs" were held on the solid ice surface. The Baltic Sea also froze over twice in the 14th Century.

The Thames Frost Fairs began in 1607 and the last one took place in 1814. It was not uncommon for the River Thames to freeze over for up to two months. During the Winter of 1683 even the seas of southern Britain were frozen solid for up to two miles from shore. Despite the famine and death resulting from the harsh winters resilient Londoners would setup shops, pubs, ice skating rinks, football pitches, bowling matches, fruit-sellers, shoemakers and even barbers on the frozen river.

The ice at the Thames Frost Fairs was thick enough for an elephant to walk on and even Kings and Queens attended the festivities. [5]

Sword, Famine, Wild beasts, and Plague

A Great Famine began in the spring of 1315. Cool wet summers brought torrential rains wrecking crops and cold dry winters killed livestock across Europe. The deterioration of all aspects of life aggravated class conflict and generated political unrest in previously prosperous countries as millions of people starved. There were reports of cannibalism.

The famine persisted until the early 1320's. It was followed by the Black Death which killed up to 75 million people with death rates ranging from 30 to 60 per cent in different areas across Europe. The effects of hunger and malnutrition and the plague were devastating. Food shortages killed millions in Europe between 1690 and 1700 and there were famines in 1725 and 1816. The final terrible misfortune of that cold period was the "hungry forties" of 1840-1850 during which the Irish Potato Famine occurred after a period of unusually cool moist weather[6].

The social effects that resulted from the appalling suffering of the people contributed to one of the most barbaric practices of modern times – the witch hunts. Pope Innocent VIII aggravated the fanaticism and superstition in 1484 when he gave credence to the superstitious belief in the existence of witches and the way they affected the weather. The hysteria increased on both sides of the Atlantic and evidence indicates that surges in European witch trials coincided with the worst weather conditions during the 16th and 17th centuries. People were desperate for an explanation of the disasters. Superstitions flourished and the most destructive misconception was the attribution of bad weather to witches. Young women with the slightest physical or mental differences or attitudes were accused of being witches with many denounced out of spite or jealousy. They were subjected to humiliating trials and killed. Often children were used to cast accusations, hysterical young girls were paraded in court screaming and contorting and testifying to things only they could see. Trial Judges were often highly educated people. In the case of the Salem Witch Trials in 1692 those assessing the evidence and delivering the verdicts were Harvard educated.

Then and Now...

The Salem Witch Trials of 1692 following the cold winter of 1691 saw a collective mass hysteria amongst young girls. They would point, accuse, experience bodily contortions and whimper and babble about spirits and demons only they could see. As author Mark Twain said; "History doesn't repeat Itself, but it often rhymes".[1]

The cold climate also influenced geopolitics.

In the first half of the 17th century, cold and dry weather in China caused famines and floods. The peasants could not pay the taxes demanded by the ruling Ming Dynasty. A peasant revolt against the weakened local authorities provided an opportunity for Manchurian invaders from the north to cross the Great Wall. They joined the rebels and established the Qing Dynasty.

In France, the last decades of the 18th century featured bad weather that caused poor cereal harvests and cattle disease. The winter of 1788/89 was especially bitter as heavy snowfalls blocked roads, major rivers froze and commercial life ground to a halt. The shortage of bread especially upset the urban poor and sparked rioting, looting and strikes that contributed to the unrest and discontent in the country before the French Revolution of 1789.

On a positive note, the Little Ice Age possibly caused the Industrial Revolution. As more trees were felled for firewood and failed to regrow in the cold climate, people sought other methods of heating. The first steam engines were designed and built to help pump water out of coal mines so that coal could be brought to the surface and used as fuel for fires. The steam engine was the catalyst of the Industrial Revolution and the original spark that resulted in the modern society we enjoy today.

The Benefits of Warming

The Holocene has seen several periods of warmer temperatures. In addition to the Medieval Warming Period there was also the Roman Warming Period (RWP) when it was warm enough for vineyards in Northern England and for Hannibal to cross the Alps with elephants in Winter[7]. Global proxy temperature estimates show that the MWP and RWP were both warmer than today by at least one degree. The cold period that punctuated The Roman and Medieval Warming was known as The Dark Ages.

There were several other periods of cooling and warming during the Holocene and the historical record tends to show the same outcomes. Research by *The Institute of Geology and Geophysics* in Beijing found that when the climate warmed, Chinese civilisation prospered and when it cooled, it declined[8]. It is common sense that food production will be higher in warmer climates, the need to find and build shelter is minimised and disease will not as easily ravage populations which are well fed and healthy. This gives time for people to settle a wider expanse of territory and then to farm, build towns and cities and spend their energy and time on innovations and inventions rather than scavenging for food or fighting wars for scarce resources. The recent warming in our times has contributed to record harvests and the ever-increasing amount of food that is being produced around the world. Far too many people are still starving but the resources are available to feed them if appropriate policies are put in place.

A warmer planet also means less energy is required for heating. This is a benefit for people who like to be comfortable in winter and prefer affordable electricity bills. Concurrently there will be a lower prevalence of illness and disease caused by cold weather. People tormented by the fear of warming need to know how much death and illness is caused by cold weather.

> Over the last five years there has been an average of 32,000 excess winter deaths in the UK every year. Of these, 9,700 die due to a cold home – the same as the number of people who die from breast or prostate cancer each year.[9]

Winter deaths are aggravated by rising prices for power. Cold homes cause easily preventable cardiovascular and respiratory illnesses. When the temperature of a room drops below 16-17^{0}C the chances of respiratory illness increase dramatically which can often be a death sentence for the elderly and very young. Indeed, during Australian winters, hospital admissions for respiratory illness amongst the elderly have increased as electricity has become more expensive. Hot weather causes discomfort but not sickness and death, provided that people are hydrated. Much of the distress caused by hot weather can be avoided by air conditioning, if it is affordable.

Economic models developed by Economics Professor Richard Tol and others indicate that the desirable effects of warming will outweigh the undesirable effects up to another two degrees of warming. That is hardly surprising in view of the benefits detailed above. Even if temperatures exceed the two degrees of warming mentioned by Professor Tol there are models that indicate that the transition will not be catastrophic. There is plenty of time to make plans. Even in the worst-case scenarios there is at least a hundred years in which we have time to adapt.

By that time, all the nations in the world will be immensely richer and advances in science and technology will surely provide us with ways to mitigate any potential warming. The benefits of warming are obvious enough especially when compared with the disastrous privation of the Little Ice Age and the extremely cold winters in many parts of the world.

Vital Signs

So, what are the problems with warming that alarmists allege? People who are panicked by the possibility of an increase in temperatures, advance a long list of issues. These are examined in detail in other chapters. Yet the reality is the list of issues are all disputed in the scientific literature. Fears are usually exacerbated by so-called "extreme weather events", which are not actually more prevalent these days, or by cherry-picking data from phenomena that go in cycles, like the melting of polar ice.

Looking at The Global Thermostat indicates clearly that if we had a choice, it would be better to turn the dial up a few degrees rather than down a few degrees. A degree or two colder and we are in real trouble, in the other direction we are safer, healthier and more prosperous.

That warmer climates meant more prosperous civilisations and colder ones saw famine and death, should not be surprising. We associate the dark and cold with decay and death whilst light and warmth are synonymous with growth and life. We are literally created by warmth. The temperature of a pregnant woman is 37.8°C, almost a degree higher than normal body temperature. Look at the places where most people go on holiday and the locals have the most relaxed lifestyle (hint: it is not Iceland or Greenland).

Humans, plants and animals all behave contrary to what alarmists claim. Life seeks and thrives in warmth; animals and plants are most prevalent and diverse at the equator where it is warmer and usually wetter. There are very few species at the poles and even fewer trees and vegetation. Appreciate the warming of the last 150 years and the greening of the planet caused by the extra plant food in the atmosphere. Do not be alarmed by the prospect of another degree or two of warming over the next hundred years.

Chapter V
THE PHANTOM MENACE
The Carbon Chimera

"It is always a much easier task to educate uneducated people than to re-educate the mis-educated."

HERBERT M. SHELTON

How much warming have we had since the industrial revolution? How much we can expect in the near future? Do we need to be concerned about the role of CO_2 as a driver of warming?

The Earth has warmed about 1°C since 1850 and there has been no warming for the last 20 years. The worst-case scenario predicts one or two degrees more by the next century or two, assuming we do not get the abrupt cold change that some scientists have predicted. As for the role of CO_2, alarmists have portrayed it as the global thermostat capable of altering the temperature of this enormous planet with its miles of atmosphere as well as a terrible pollutant and a threat to all life on Earth. First, let us address what the history of the Earth has to say about this.

The Long-Term Geological View

If CO_2 were the primary driver of warming as the alarmists claim, the half-billion-year reconstruction of temperatures and CO_2 levels in the chart below should show temperature and CO_2 moving in lockstep, yet it shows nothing of the sort. The period marked A (315 – 270 mya) is the only time in the last 600 million years when CO_2 and temperatures were as low as they are today (B).[1]

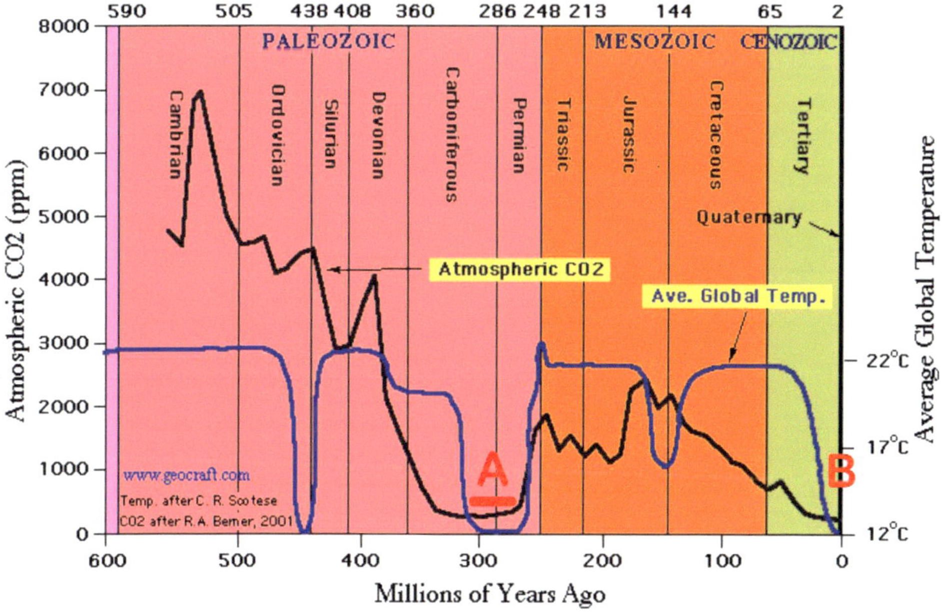

The second chart of just the past 60 years shows that the steady and continuous rise of CO_2 is mostly out of step with the temperature except a short period in the 1980's and 90's. This too suggests that there is no correlation or causal relationship between CO_2 and warming[2].

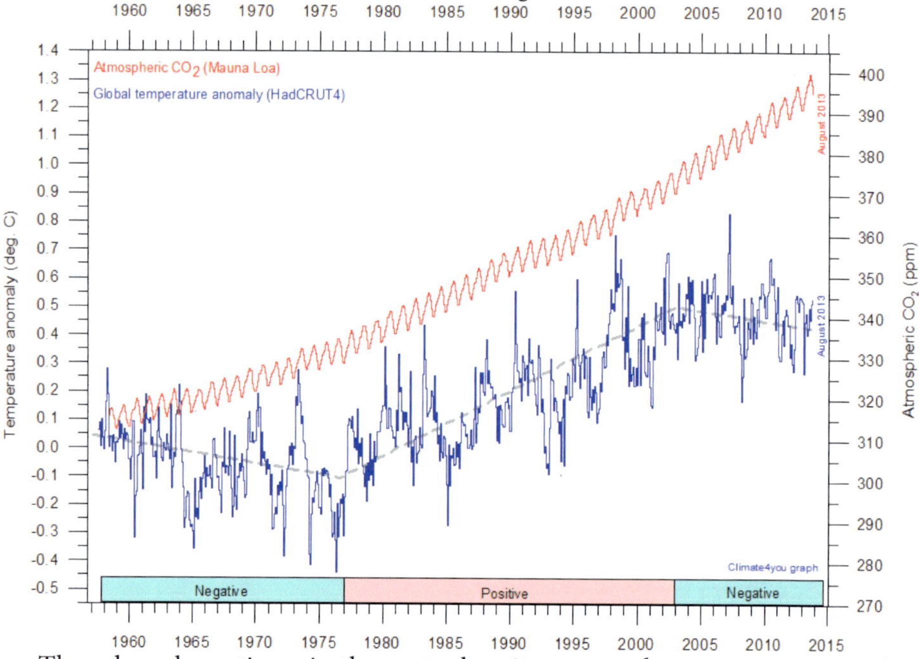

There have been times in the past when it appears that temperatures and CO_2 have increased together. This is nothing unusual if you think about it. Temperature is either going up or going down, likewise CO_2. It is inevitable that at some point they are going to be increasing or decreasing together. For the theory of CO_2 as the principal temperature control for the Earth to hold then temperature and CO_2 must go up and down together (or even with a small but consistent time lag between them) always and forever. As soon as temperature and CO_2 do not correlate then the theory of CO_2 as the driver of warming fails. If Isaac Newton dropped a ball 999 times but on the 1000th experiment the ball shot up into the air never to be seen again, he could never have written the laws of gravity.

The period when CO_2 and warming were in step is the time when the foundations of the great climate computer models were laid in the early days of alarm. The models are probably the most important tools used to support the IPCC reports and the correlation of CO_2 and warming during that short time appeared to support concerns. When the pause or hiatus in warming arrived after 1998 the predictions based on the models drifted further from the measured temperature every year.

The danger of excessive dependence on computer models is that, rather than admit the models are faulty, some scientists will simply ignore reality. CO_2 was put forward as the driver of temperature far too early. There was not enough research or data to justify that conclusion. This is still the case despite the trillions of dollars that have been spent on climate science.

The hockey stick that pushed the alarm

The most famous, now thoroughly debunked, alarmist paper was written by Michael Mann and associates at the East Anglia University Climate Research Unit (CRU). It inexplicably cooled the Medieval Warming Period and spliced on a different set of data to the end of the graph to generate a "hockey stick" showing a steep rise to "unprecedented" levels in recent times. What this paper essentially said is that temperatures today are warmer than the Medieval Warming Period (950-1250 AD) when the Vikings sailed to Greenland, established settlements, grew crops and buried their dead. Records indicate the settlements were abandoned around the 15th Century as the World cooled. Today, owing to the permafrost, the ground is frozen solid and only industrial machines can excavate Viking artefacts. Most of Greenland is covered in an ice sheet 2-3km thick and is pretty easy to spot from space as it permanently white. The Vikings decided to call this place Greenland. Think about that; **Green**land, yet according to climate alarmists and their settled science, when the Vikings colonised and lived there it was **colder** than today.

The hockey stick graph was a public relations dream for the IPCC and it provided the background for the findings of their Third Assessment Report in 2001. After the graph was shown to be a false alarm based on some highly selective sampling and defective statistical methods the IPCC quietly parked it in the archives as though it never happened.

Mann, however, insisted his graph was legitimate and anyone who dared criticise it was subjected to lawsuits, though when Mann took Canadian commentator Mark Steyn to court to defend the credibility of his hockey stick graph not a single climate scientist was prepared to join Mann in the lawsuit.

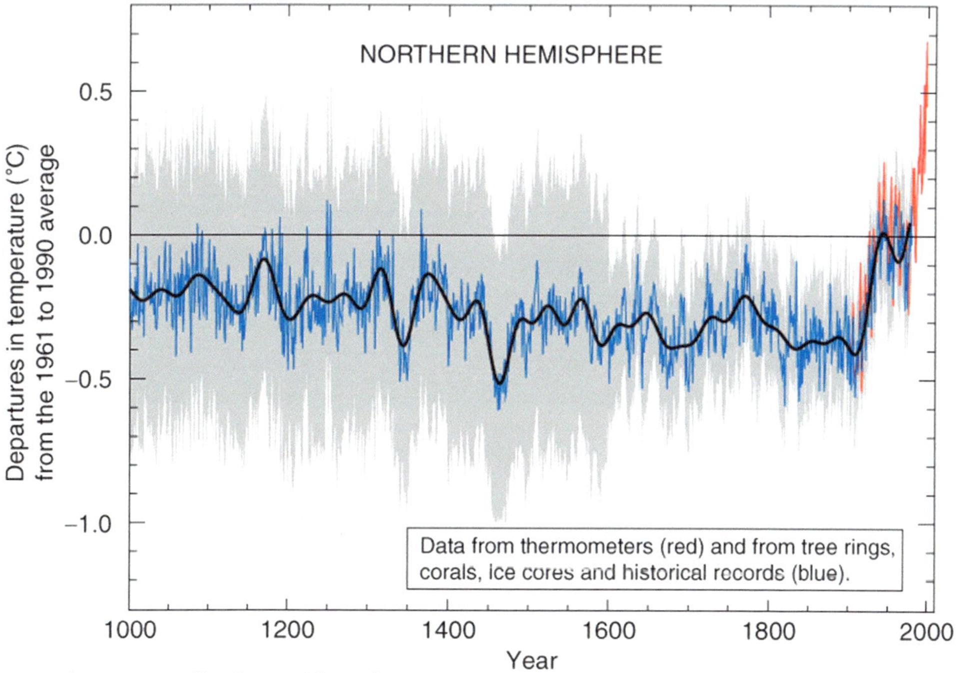

The notorious "hockey stick" graph. Prominent in the 2001 IPCC report, now forgotten.[3]

In August 2019 Mann lost a court action against Canadian scientist and "denier" Tim Ball. Mann could not or would not produce the data he used to create his infamous hockey stick graph and so the Judge awarded the case to Tim Ball along with significant damages.

In nearly every case brought before the Courts the alarmists have lost, usually because they could not or would not produce evidence. The "deniers" have won every time they have been sued or challenged on the evidence and this is despite having no funding compared to the billions of dollars of taxpayers funds that the alarmists have access to through Universities and Foundations.

Tim Balls' win was one more win to the "science deniers", whilst the alarmists still have nothing on the scoreboard.

Al Gore also used the hockey stick graph as his primary source to claim temperatures were rising dangerously in his documentary *An Inconvenient Truth*. Gore's movie was riddled with inaccuracies, false statements and deceptive propaganda which was confirmed in a UK High Court case in 2007[4]. Not only did he use the unscientific hockey stick graph, he used a second graph with a scale of just 650,000 years to compare warming and CO_2 and emphasised the scare tactics by mounting a scissor lift.

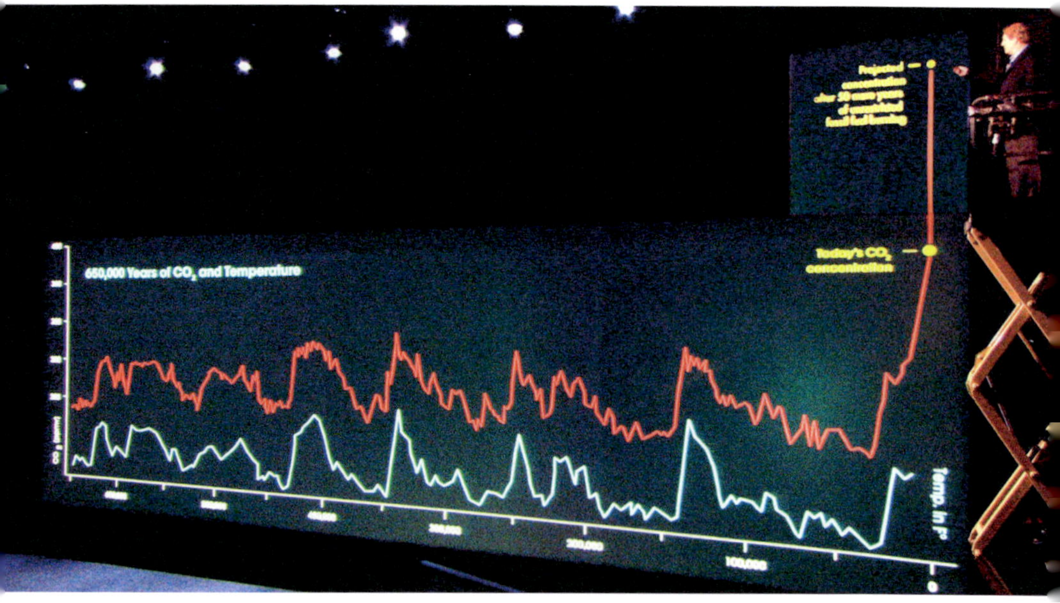

650,000 years may sound like a long time but if you look on the "Global Temperature and Atmospheric CO_2 Over Time" graph (the first graph in this chapter), that time scale is smaller than the thick black line on the far right of the graph which itself only represents 600,000,000 years of the 4,500,000,000 history of the Earth. Or in other words, Gore used 0.1% of the graph whilst ignoring the other 99.9% of inconvenient truth which shows no correlation between CO_2 and temperature.

The Gore scissor lift graph is pulled out frequently by alarmists of "proof" that CO_2 drives temperature, yet that is completely backward. When you look closer, or zoom in, you see that the rise in CO_2 clearly follows the rise in temperature. The three graphs below[5] show how the Gore deception worked. The top graph is the one going back 600 million years, the middle graph is the Gore graph stretching back just half a million (500,000) years and the bottom graph zooms into 50,000 years. Gore just found a convenient angle that appeared to show a correlation where none existed. He should have mounted a cherry picker rather than a scissor lift.

Why do increases in CO_2 follow a rise in temperature and decreases in CO_2 follow a fall in temperature? On very short timescales of a few hundreds or thousands of years, when the temperature of the oceans increase they de-gas, meaning CO_2 is released. This is exactly what happens in a soda bottle in the sun, it goes flat quickly because the warm soda releases CO_2. If we keep the soda in the fridge it stays fizzy for much longer since CO_2 is more soluble in cold liquid. The oceans are somewhat larger than a bottle of soda and thus this process takes much longer, in the hundreds and thousands of years. Some scientists have postulated that the increase in natural CO_2 today is as a result of the Medieval Warming that began over 1000 years ago and lasted for 300 years or perhaps the earlier Roman Warming Period.

The Bathroom Effect

If more CO_2 in the air could raise the temperature of the oceans then it would be very easy to warm a bath. Simply fill the bath will cold water, add some extra CO_2 into the bathroom (maybe set off a fire extinguisher) and hey presto! A warm bath! Think of all the money you will save on your power bills. Even if you pointed a bar heater at the water in the bath, how long and how much energy do you think the water would require to warm just 1°C?

Here is another thought experiment – what happens when you fill a bath with hot water in a cold bathroom? After a while the temperature of the bathroom will be raised and the bath water will still be warm enough to take a bath. This should tell you everything you need to know about what the primary control of the climate is.

Climate change is fluid mechanics (the oceans) and radiative physics (the sun), the trace gas making up just 0.04% of our atmosphere has no perceptible effect on the temperature. Despite what alarmists claim CO_2 is not a magical heating gas.

Even if CO_2 and temperature were aligned or highly correlated this proves nothing since one of the major lessons of science is that correlation does not equal causation. Alignment or correlation on a chart can happen by coincidence when there is no cause or driving effect from one thing to the other. For example, among children there is a correlation between the size of their feet and their general knowledge. Yet, feet growth does not cause you to become smarter. There is also a correlation between warm weather and hay fever, yet it is not warm weather that causes hay fever, pollen does that. Someone suffering from hay fever would be symptom-free in the very hot Sahara Desert yet in the cooler tulip fields of Holland they would be suffering immensely.

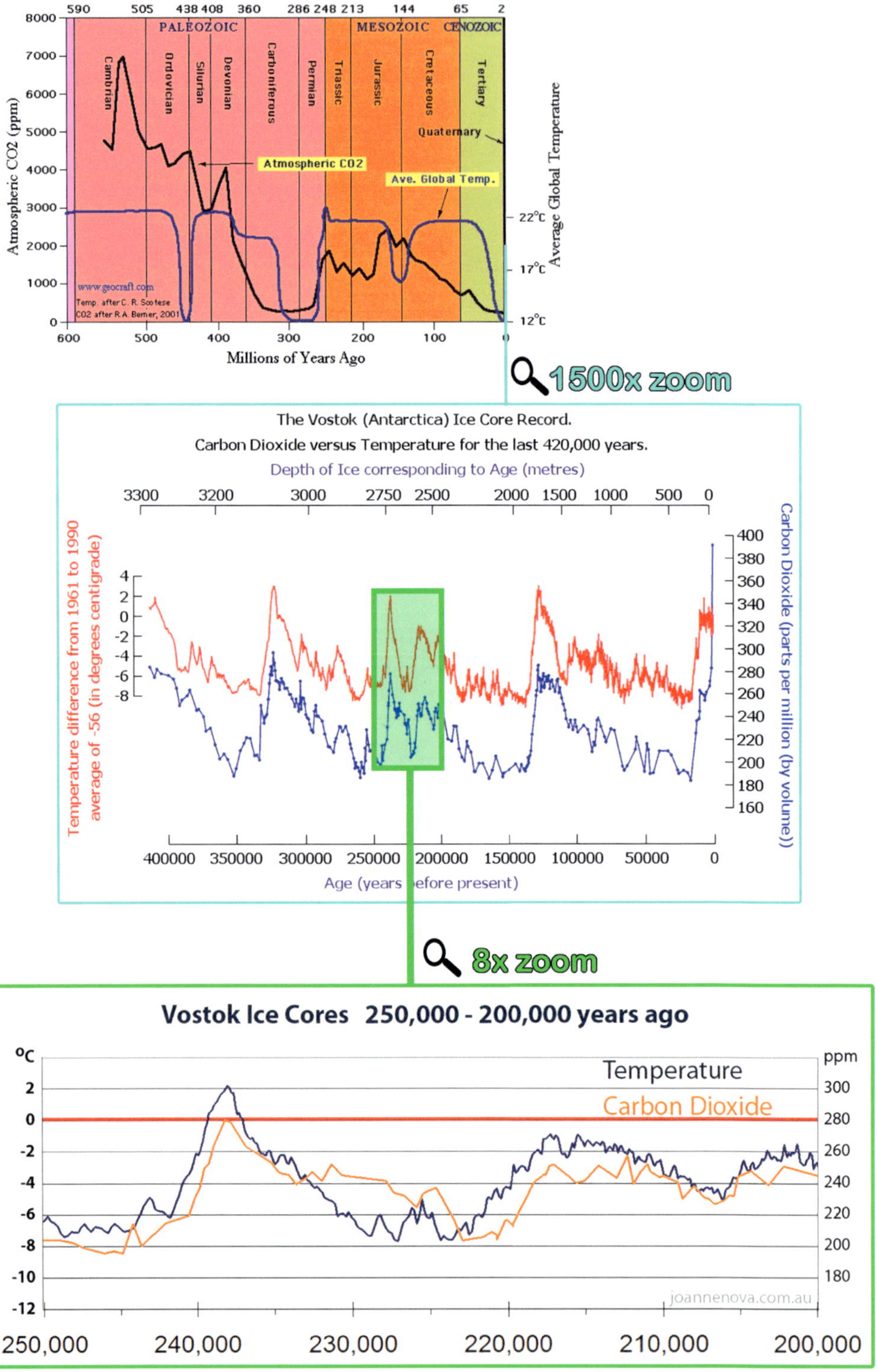
PALEOZOIC
MESOZOIC
CENOZOIC
Cambrian
Ordovician
Silurian
Devonian
Carboniferous
Permian
Triassic
Jurassic
Cretaceous
Tertiary
Quaternary
Atmospheric CO2
Ave. Global Temp.
Atmospheric CO2 (ppm)
Average Global Temperature
Millions of Years Ago
www.geocraft.com
Temp. after C. R. Scotese
CO2 after R.A. Berner, 2001
1500x zoom
The Vostok (Antarctica) Ice Core Record.
Carbon Dioxide versus Temperature for the last 420,000 years.
Depth of Ice corresponding to Age (metres)
Temperature difference from 1961 to 1990 average of -56 (in degrees centigrade)
Carbon Dioxide (parts per million (by volume))
Age (years before present)
8x zoom
Vostok Ice Cores 250,000 - 200,000 years ago
Temperature
Carbon Dioxide
joannenova.com.au

A Trick of the Eye

We have seen how graphs can be made to look more dramatic through various methods. Zooming into a certain time period, producing charts which only show a certain trend, cherry picking a particular geographic area that shows what they want you to see but is just a snapshot of a much bigger picture, these are just some of the tricks alarmists use, with alarming success. Let's look at some further examples by looking at graphs below of temperature from the past 100 years and 15,000 years.

It is strange how the alarmists only show you certain time scales but ignore or hide others if they don't show what they want you to see. In the past 10,000 years (the age of civilizations) the temperature has warmed and cooled at faster rates than today. Yet the IPCC and the alarmists fret over a single degree of warming. That change is negligible compared with the variations that we take in our stride from hour to hour through the day, between day and night, between the seasons and between different parts of the world. Those living in Melbourne, Australia would be able to appreciate this! In one day in 1972, the temperature in Loma, Montana changed by an astonishing 57.2°C (103°F).

Alarmists reinforce their claims of a runaway warming effect of CO_2 by referring to a "forcing factor" or "feedbacks". That is a kind of unintended knock-on effect that more CO_2 will have on the atmosphere. Yet this runaway factor has never been seen or measured before and the historical climate record shows absolutely no evidence for it. As such it only exists in theory inside computer models, which have failed, and the minds of alarmist scientists, which have been consistently wrong in their predictions. These apparent "feedbacks" are very alarming to physicists as they tend to break the laws of physics, specifically the first law of thermodynamics.

Beware of The Models!

Recent investigations suggest that the warming effect of CO_2 is only a fraction of what the IPCC thinks it is. It seems that the assumptions about the effect of CO_2 built into the most advanced models used by the IPCC are based on speculations from over a century ago. In 1896 Swedish Professor Svante Arrhenius postulated that doubling atmospheric CO_2 would increase the global temperature by 5 to 6°C. This was based on extrapolation of some early attempts to measure the surface temperature of the Moon and highly speculative estimates of the infrared radiation leaving the Moon and reaching the Earth. Arrhenius himself wrote that it was necessary to obtain experimental evidence for his predictions but no such equipment existed at the time to perform these experiments and Arrhenius himself recognised that he could very well be proven wrong. Indeed, when scientific equipment was invented and experiments were carried out some years later it was discovered the Arrhenius was wrong by a significant factor. He had vastly overestimated the amount of warming, miscalculated the logarithmic effect of CO_2 and missed the crucial narrow absorption bands of the gas. What this means is that any warming from CO_2 happens up to around 100ppm and after that it doesn't matter how much CO_2 you add, it just isn't going to warm very much. It's like a sponge that is completely soaked, no matter how much water you pour on it the sponge just isn't going to absorb any more water.

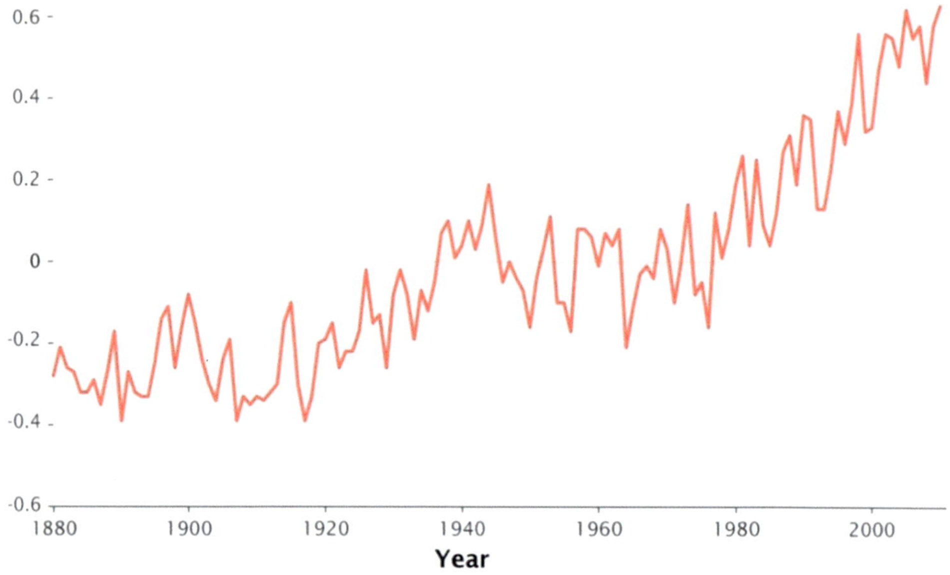

The graph above[6] may look alarming at first but when you look at it in the context of past warming it is entirely unremarkable[7].

The red line at the end of the Little Ice Age is the temperature increase shown in the "alarming" graph.

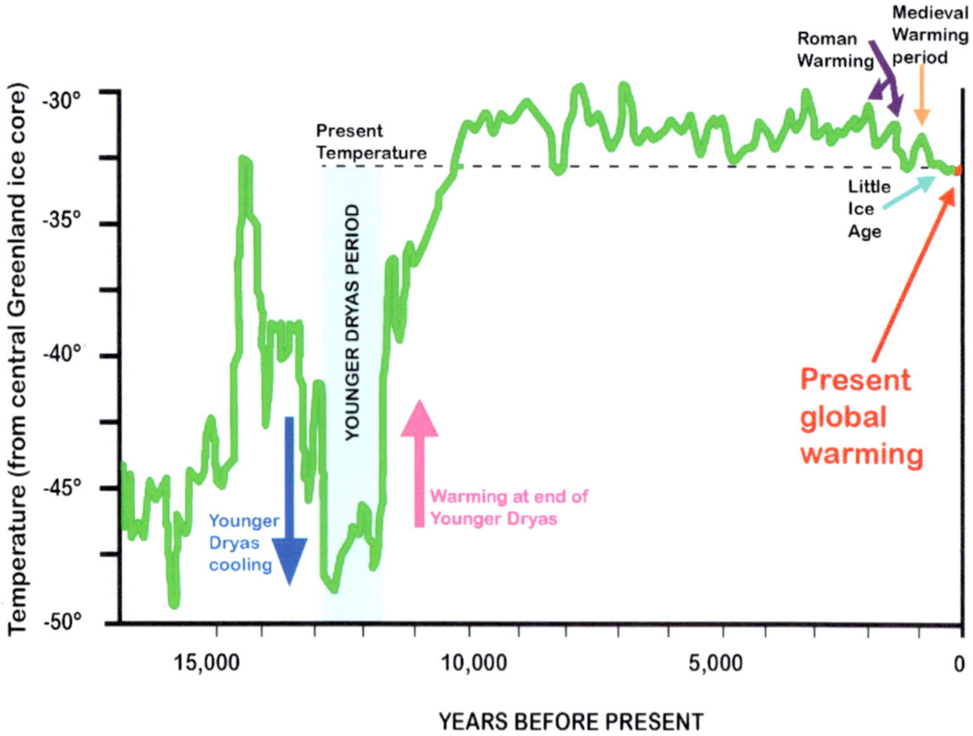

When alarmists refer to "experiments carried out 100 years ago prove this" they never refer to the inconvenient truth that Arrhenius vastly overestimated his findings, that he feared this was the case and that he actually thought an increase in CO_2 would be a good thing!

> By the influence of the increasing percentage of carbonic acid in the atmosphere, we may hope to enjoy ages with more equable and better climates, especially as regards the colder regions of the earth, ages when the earth will bring forth much more abundant crops than at present, for the benefit of rapidly propagating mankind.[8]

Arrhenius' speculations became public and important when NASA's James Hansen detonated the warming scare as a political issue with a presentation to a U.S. Congressional Committee in 1988. The event was held on a very hot summer day with the air conditioners in the chamber deliberately sabotaged to heighten the effect and so the TV cameras would have hot, sweaty scientists to focus on as they told the world how the world was heating up dangerously. The participants later bragged about their sabotage. Ask yourself, if the evidence was so compelling then why the need to manufacture and manipulate the public? The Global Warming invention began in the way it has been maintained to this day, with manipulated temperatures, visual propaganda tricks and shonky computer models. Hansen based his alarming presentation on the results that were coming out of the computer models that were programmed with versions of the original Arrhenius calculations. A key point was his assertion that doubling CO_2 would warm the world by 4.2°C.

The IPCC began by nominating the largest number that they thought they could "sell" to the scientific community at large and that number was a compromise between the more gung-ho IPCC scientists and those who were inclined to be more circumspect. The IPCC then revised it downward in small stages. Abbot and Nicol[9] noted the adjustments from 3.8°C in 1995 to 3.5°C in 2001 and 3.26°C in 2007. Remarkably they report a statement by Hansen himself that it might need to be reduced further to 2.5°C. Other investigators have been bolder in downward adjustments, some are as low as 1°C and others claim that the suppressing effect of CO_2 could outweigh any forcing effect and so there is likely no net effect at all.

The Miracle Molecule

Alarmists believe CO_2 at 400ppm is something to be worried about. It is difficult for most people to comprehend 400ppm or 0.04% of the atmosphere. So, let's use a stadium the size of the Rose Bowl in Los Angeles or the Melbourne Cricket Ground which can seat 100,000 people[10]. If a packed stadium represented the atmosphere CO_2 would be represented by just 40 people and of those 40 the contribution of man-made CO_2 would be equal to approximately one large man. Can we really believe that one man could influence the atmosphere of an entire stadium of 100,000 people? Even the loudest, drunkest sports fans cannot do that.

Even though CO_2 is understood to be a GHG this does not enhance its insignificant role. Of all the GHG's water vapour is by far the most important. Climate scientists estimate it to cause 50% of warming in the atmosphere, clouds another 25% and CO_2 at 20% (and that is the high end of the estimate).

The Atmosphere in the Stadium

A stadium of 100,000 people:

78,000 people are Nitrogen (N_2)

21,000 people are Oxygen (O_2)

930 people are Argon (Ar)

3 people are trace gases;
Neon (Ne)
Helium (He)
Methane (CH_4)
Krypton (Kr)

Just 40 people are natural carbon dioxide (CO_2)

Man-made CO_2 is...ONE (slightly overweight) person of the 100,000 people in the stadium

Note: Figures based on a dry atmosphere, water vapor (the main GHG) is not included. Water vapor is about 0.25% by mass of the full atmosphere.

Methane, ozone and aerosols account for the rest. According to the theory, without GHG's the Earth's temperature would be 33°C cooler[11] (about -18 °C). Thus CO_2, at the high end of estimates, is responsible for about 6.6°C of warming of which, by the IPCC's own admission, 95% is natural leaving just 5% man-made. This means that mankind, globally, is "responsible" for no more than 0.33°C of warming (5% of 6.6). One third of one degree. Australia accounts for 1.5% of emissions globally and if it ceased all CO_2 emissions tomorrow it might reduce the temperature by a massive (1.5% of 0.33) 0.005 degrees, which is five thousandths of one degree.

CO_2 has increased from 280ppm at the beginning of the Industrial Revolution to the average of around 400ppm today. Alarmists do not tell people that most CO_2 is natural and not man-made. It is estimated that mankind is responsible for just 3 to 4% of emissions; 97% of all current atmospheric CO_2 derives from natural sources. We know this because it is possible to measure natural CO_2 and CO_2 produced by the burning of coal, oil and gas. However, so-called "man-made CO_2" is no different to natural CO_2, it does not hang around in the atmosphere forever but flows through it into plants, rocks, the seas. This is known as the carbonate-silicate cycle and CO_2 takes about four years before it is recycled through natural processes.

Even the 3% estimate of man-made CO_2 is considered to be an overestimate by some scientists. The IPCC does not include the huge volumes of natural CO_2 emitted from the millions of submarine volcanoes and vents in their calculations. This natural CO_2 has the same fingerprint as burning hydrocarbons. How do we know the volumes, time periods and processes taken for this natural CO_2 from deep-sea volcanoes to be released into the atmosphere? We don't as it has never been measured, yet we are told the science is settled.

Regardless, even taking the worst assumptions and faulty science of the IPCC, their claims that mankind "adds" 1.5ppm of CO_2 each year and assuming it is not absorbed by plants and oceans (which it is) then it would take more than 250 years to go from today's 400ppm to 800ppm. Imagine telling the MCG or Rose Bowl that they should expect an additional 40 fans over the next 250 years, do you think they would have enough time to plan? In reality with this massive lead time we have plenty of time to invent and innovate, 250 years ago nobody could even have imagined telephones, microwave ovens or Netflix, Captain Cook had only just sighted Australia. Can you imagine where we will be in 250 years?

And what happens at 800ppm? Well as horrific as that level sounds and as much as the alarmists bleat about mankind never experiencing these levels of CO_2, 800ppm is not unusual in the least. Whilst the alarmists are technically correct it is not true that man has never experienced 800ppm CO_2, in fact we do so nearly every day in the office, the classroom, the car and even outside. Submariners and astronauts are breathing CO_2 at concentrations of 4000-5000ppm as you can see in the diagram[12].

CO_2 Parts Per Million

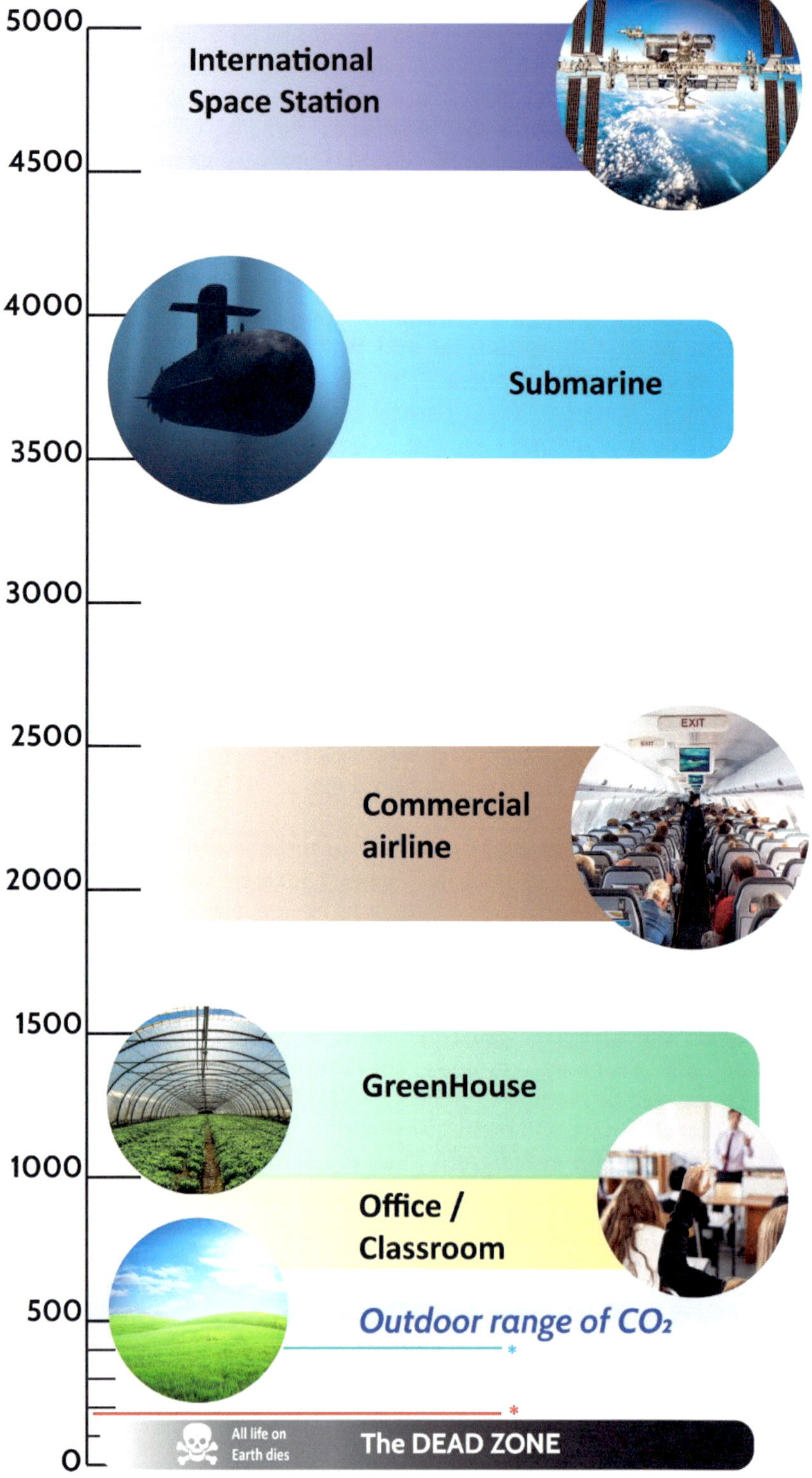

* Mauna Loa, Hawaii (~410ppm)

* Lowest level of CO_2 in Earth's history

CO_2 levels are not "fixed" at 400ppm, it is not a global constant. Depending on the time of day, the season and where you are in the World, outdoor ground level CO_2 levels can fluctuate from as low as 200ppm to nearly 1000ppm. You can measure this yourself easily with a cheap CO_2 sensor. Try it in a crop field in Summer during the day and the night.

CO_2 is Greening the Planet

Whilst warming has been good, extra CO_2 in the air is even better. Carbon is the basis of life on earth and the CO_2 in the atmosphere is an absolute requirement for plant growth. Moreover, the level of CO_2 at present is only just enough for green plants to survive and far below what they need to thrive, our plants are literally gasping for breath. In commercial glasshouses the level is elevated to four or five times the concentration outside. Some corals thrive with six times the normal concentration of CO_2 in the water.

The value of additional CO_2 has been recognized for some time and there is a lot of scientific literature on the topic. This has recently been reported by the CSIRO who are often cited as a source of truth by the alarmists.

> Increased levels of carbon dioxide (CO_2) have helped boost green foliage across the world's arid regions over the past 30 years through a process called CO_2 fertilisation, according to CSIRO research.[13]

British journalist Matt Ridley explained that this should not come as a surprise. He reported a study published in 2016 with 32 authors in 24 research centres based in eight countries. Satellite data collected over 30 years showed a 14% increase in green vegetation with three quarters of the increase due to the extra carbon dioxide in the atmosphere.

> Global greening has affected all ecosystems – from arctic tundra to coral reefs to plankton to tropical rain forests – but shows up most strongly in arid places like the Sahel region of Africa, where desertification has largely now reversed.

Higher CO_2 makes plants more drought resistant as their stomata (the plants mouths) are not having to gasp for CO_2 and lose water in the process. This is great for a country like Australia as less water is required and it also makes crops cheaper to grow.

Climate alarmism promotes the idea that CO_2 is a pollutant yet as can be seen quite clearly from the diagram, people are frequently breathing CO_2 at much higher concentrations than outside. Even when we are exhaling, our breath is 40,000ppm (4%) CO_2, which is mostly the "exhaust" from the food we consume. If you think CO_2 is a pollutant then you should stop breathing.

Furthermore, the Earth has historically seen much higher levels of CO_2 and there is no evidence of catastrophic weather, boiling seas, mass extinctions (at least not caused by CO_2) or any other of the litany of crazy predictions and apocalyptic scenarios advanced by alarmists. *Au contraire*, higher CO_2 levels are generally associated with life and growth whilst in the past when CO_2 was lower humanity not only suffered considerably, we were nearly wiped out as a species. At around 180ppm life on Earth begins to die out as nothing can grow and sustain life.

What level of CO_2 would you like?
Imagine the volume control on a speaker. Imagine a similar control globally and for CO_2 rather than volume. This is The Carbon Dioxide Dial[14] that you see in the diagram. Look at where we are now at 400ppm and notice how close we are to the 'Dead Zone' where all life on Earth dies out because there is not enough plant food (CO_2) for plants and trees to grow and thus no food source for animals or humans. Also notice how when CO_2 was at much higher levels than now there have been Ice Ages.
Now ask yourself whether you would prefer;

A) A reduction of 300ppm CO_2?

B) An additional 300ppm CO_2?

If you answered A) then you have condemned the Earth and every person, animal, tree and plant to death.

Answer B) and life on Earth will flourish.

Do you still think CO_2 is dangerous? A pollutant? A potent gas capable of heating? How come all those bubbles of CO_2 don't turn your beer warm or greenhouse your coke? Why do we use CO_2 in fire extinguishers to put out fires? The solid form of CO_2 is that rather fun substance called dry ice that is used mostly for refrigeration and preserving frozen foods. Have you ever seen a CO_2 filled blanket in a shop? Or a winter jacket infused with CO_2 so it can trap more heat? Maybe we could start selling these to gullible global warming believers and become millionaires like Al Gore.

Some alarmists have spoken of machines or techniques that could suck CO_2 out of the atmosphere. This would be the equivalent of a doomsday device only seen in fictional movies about alien invasions and be far worse than atomic weapons. Such a machine could potentially wipe out every living organism on the planet rendering it dead and lifeless. Even reducing CO_2 slightly might end up triggering a runaway cooling where the oceans re-gas, pulling more CO_2 from the air. Colder weather causes more glaciation that reflects the sun's heat driving the sea temperatures lower and so on. The planet was very close to this scenario just 20,000 years ago when temperatures and CO_2 were on the brink of causing a planetary mass extinction. Mile-thick glaciers and ice sheets covered the Northern Hemisphere burying trees and shrubs and the cold and lack of plant food (CO_2) severely hampered plants growth. Only an estimated 100,000 humans remained, too **little** CO_2 in the air almost eradicated mankind. Look at The Carbon Dioxide Dial again and see how close we are to this extinction level catastrophe occurring again.

The Carbon Dioxide Dial

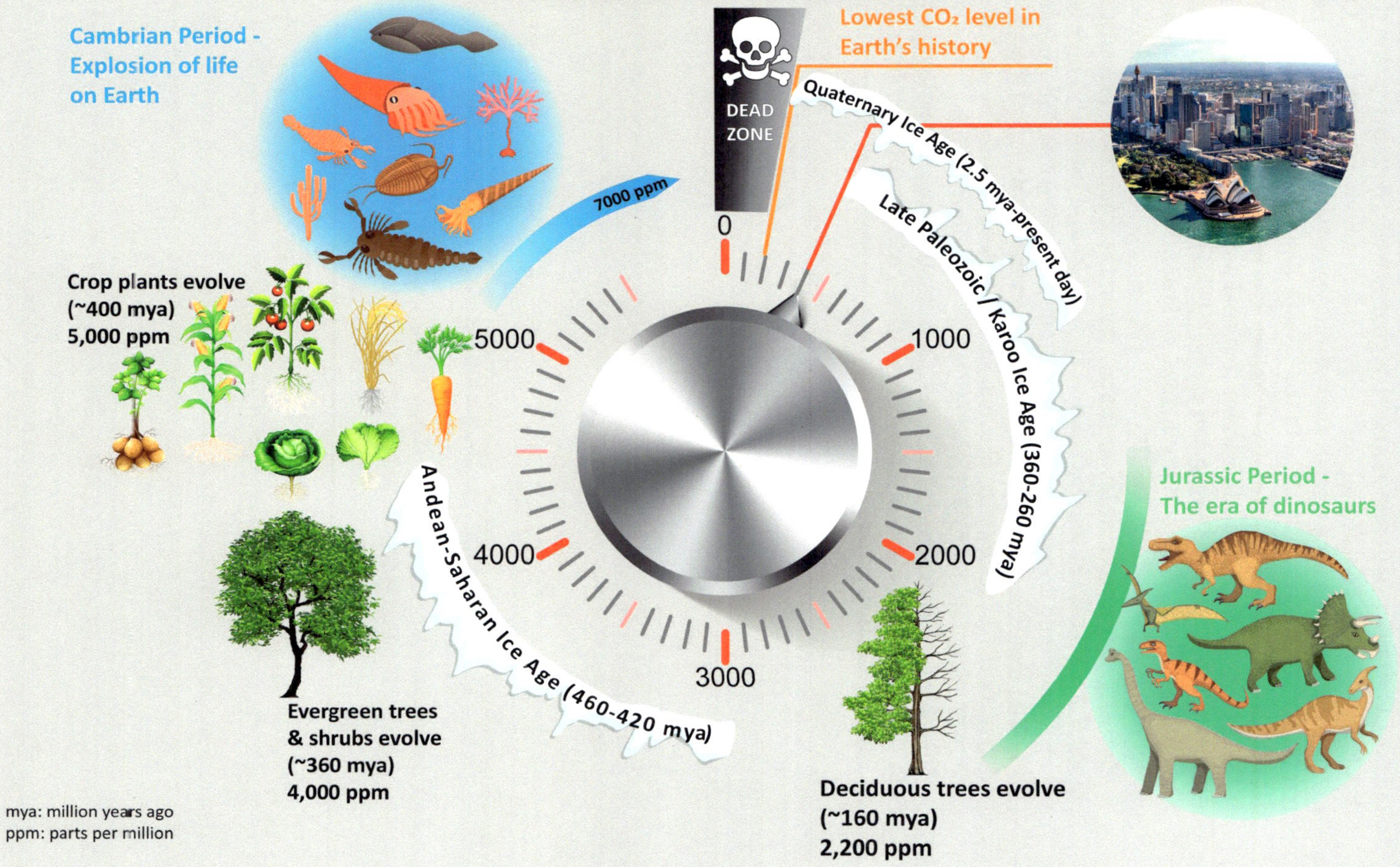

mya: million years ago
ppm: parts per million

Puny Humans

Life on earth flourished when the temperature was much higher than it is today and the CO_2 content of the air was much higher as well. That observation alone should dismiss any concerns about the present situation. The level of CO_2 is still recovering from a time when it was only just enough to sustain life on Earth and we are still dangerously close to the "Death Zone". In fact, it is far more dangerous for CO_2 to drop than for it to rise. The panic over the tiny amount of CO_2 produced by man from combusting coal and oil, the bedrocks of our modern lives, is vastly overblown and exaggerated. Humans do not and cannot control the climate. We can impact local environments to an extent but we cannot influence the global climate.

The 2011 earthquake in Japan shifted the entire country 2.4 metres and pushed the planet on its axis by nearly 4 inches (10 centimeters). There is no way mankind could do anything like this, it took nature less than five minutes[15]. This notion that humans have control over the weather, droughts, floods, sea level rise, ocean heat content, glaciers, or extinctions simply by emitting a fraction of a fraction of CO_2 is elevating us way above our station.

This chapter has demonstrated that CO_2 is not the magical heating gas that alarmists claim. The historical record proves this, physics proves it, observation and experimentation proves it.

Chapter VI
JOURNEY TO THE CENTRE OF THE EARTH
What Actually Regulates the Climate

"Your assumptions are your windows on the world. Scrub them off every once in a while, or the light won't come in."

ISAAC ASIMOV

The assumption that CO_2 is the climate control knob has been promoted so effectively that hardly anyone is aware of the factors that have a far greater impact on climate. We have been led to believe that the primary control variable of climate is an invisible trace gas, comprising just 0.04% of the atmosphere, an amount so small that it is measured in parts per million, currently a little over 400ppm on average. The idea that this trace gas, contained within just a few miles of the troposphere and fluctuating by a fraction of a percent could fry the planet is an idea that has crossed the border into magical thinking.

The Greenhouse Gas (GHG) theory is not a theory, it is a hypothesis. In the realms of science, a hypothesis sits below a theory. It is the beginnings or basics of a theory, an idea that is yet to pass enough tests to count as a credible explanation of anything.

Of course, that raises the question "*If it's not CO_2, what is it then?*" This chapter will describe the primary drivers of climate and weather to explain why CO_2 is *not* the climate control knob.

The Sun

It is relatively easy to explain climate change; it is the Sun (heat) and water in its three phases (solid, liquid and gas). Or in more technical terms it is radiative physics and fluid mechanics.

That's it.

Every so often a volcano might blow its top. This can dramatically alter the climate for a while. And there is the occasional asteroid strike and fluctuations in cosmic rays, but beyond that, nothing else, other than the sun and water, determines climate in any significant way.

The Sun is the most important of these two factors since without it water would be frozen solid and the solar system devoid of life as the "solar" part would be missing. The Sun, by volume, is 1,300,000 times larger than the Earth, is so dense at its core that a tablespoon of it would weigh about 2kg and it outputs a million times more energy in one second than humans use in a year, though only a bit of this energy reaches Earth.

The Sun looks like a regular orange ball to us, but in fact its energy output is not constant. It has cycles of activity that are determined by several factors that are still under investigation. Some of the signs of cyclic activity such as sunspots were visible to astronomers centuries ago. Sunspots appear on the surface of the Sun quite regularly, in cycles of 11 years during which the poles flip. At the beginning of these solar cycles, the solar minimum, the Sun has the least sunspots. There are several other solar cycles that occur over longer periods.

The Sun also experiences tsunamis and eruptions, known as solar flares and coronal mass ejections (CME). These can have significant effects on the Earth. In March 1989, a CME caused serious issues in the North American power grid and even knocked out power in Quebec for nine hours[1]. They are largely unpredictable.

The amount of radiation from the Sun hitting the Earth is modified by the protective shield that surrounds the planet. This magnetosphere is created by the molten iron and nickel in the core of the Earth as it spins. Like the Sun, it is not uniform in behaviour and can be stronger, or weaker over time and depending on location. The most well-known and visible aspect of this is close to the poles where the solar wind interacts at the weakest parts of the magnetosphere creating the aurora borealis (Northern Lights) and aurora australis (Southern Lights). A massive CME in 1859 meant these phenomena were seen as far north as Queensland in the Southern hemisphere and as far south as the Caribbean in the Northern hemisphere[2]. It ignited fires, fledgling telegraph communication systems failed and night was turned into day. Should such an event occur today it would cause trillions of dollars of damage and cause a global catastrophe that would take decades to recover from. Yet we are told that a trace gas comprising 400ppm of the atmosphere is the greatest threat to mankind.

When the Sun is generally behaving itself, most of the energy that reaches the Earth is reflected by clouds and some makes it to the surface of the planet. More on this later.

The Earth

Most people would have seen an orrery[3] (often called a planetarium) in a museum or on television. It is a mechanical model of the solar system used to illustrate the rotation and relative speed of the planets around the Sun. An orrery has some uses and looks very pretty but they do not accurately show the relative sizes of the planets, the distances between them, nor their elliptical orbits.

Whereas an orrery shows circular orbits in a vastly truncated form, the real world is much more complex than the simplified model that is presented to people whether for convenience or deception. The Earth's annual journey around the Sun is not a neat circular orbit of a horizontally spinning perfectly spherical ball. It is wobbly, tilty and eccentric. The Earth is fat (not flat) and tilted over between 21.5 and 24.5 degrees as it wobbles about on its axis and it orbits the Sun (according to the orthodox "science denier" Galileo) in an oval or ellipse, not a circle.

This is by far the greatest driver of climate and we experience it every year in the seasons and in longer cycles measured in decades, centuries and millennia that we cannot experience in our relatively short lifespan.

The orthodox science of climate alarmism seemingly ignores the concept of the fat Earth, as though heat is distributed evenly across the globe. The poles are often frozen mainly because they receive less energy from the Sun than the equator owing to the tilt and wobble effect. Further, and seemingly paradoxically, CO_2 actually has a cooling effect at the poles owing to the lower levels of water vapour. Yet alarmists often claim the poles will warm at the same rate as the rest of the planet. This could only happen if the Earth were flat, perhaps this is what alarmists believe.

The wobble (axial precession) is caused by the gravitational force from the Sun, Moon and other planets and changes in cycles of 23,000 years.

The tilting of the Earth accounts for the seasons. Depending on the hemisphere, when it is tilted (obliquity) toward the Sun it is Summer and when it is tilted away is Winter. If there were no tilt, there would be no recognisable seasons. The tilt changes by 3 degrees and back again approximately every 40,000 years.

Lastly, the eccentricity of the elliptical orbit of the Earth around the Sun goes from less elliptical to more and back to less in 96,000-year cycles.[4]

Milankovich cycles

The change in the wobble, tilt and orbit of the Sun could be considered rather like the "seasons" of Spring, Summer, Autumn and Winter although they are much too long to be perceptible in our lifetimes. But they do exist, they are measurable and do affect climate.

Serbian astrophysicist Milutin Milankovitch (1879-1958) developed this theory over a period of 30 years, performing his calculations by hand because there were no computers, let alone computer models in those days. Milankovitch used his mathematical model to theorise the advance and retreat of glaciers during Ice Ages and Interglacial periods, going back 600,000 years[5]. Now known as the *Milankovitch Cycle,* it was largely ignored for decades until sediment cores proved that Milankovitch was largely correct. The theory is now accepted by numerous organisations such as NASA and the U.S. National Academy of Sciences[6].

The Milankovitch Cycle correlates far more strongly to the cycles of Ice Ages and interglacials than the fluctuations in CO_2. Of course, the climate is influenced by other factors such as vulcanism and asteroids but they do not disprove Milankovitch Cycles.

Axial precession

The Earth wobbles like a spinning top in 23,000 year cycles

Obliquity

The Earth is tilted, changing by 3 degrees in 40,000 year cycles.

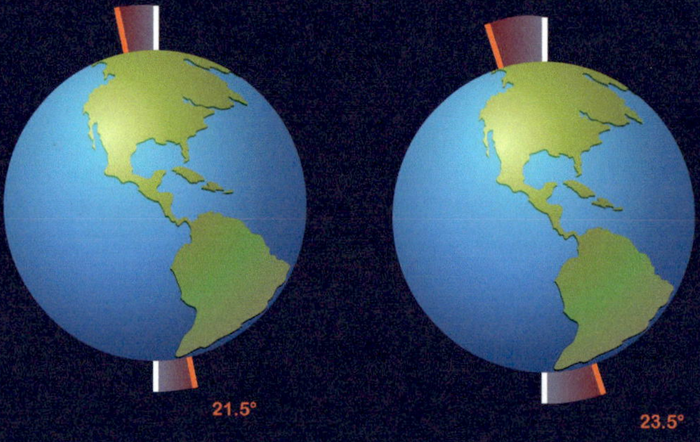

Eccentricity

The orbit of the Earth around the Sun changes in 96,000 year cycles

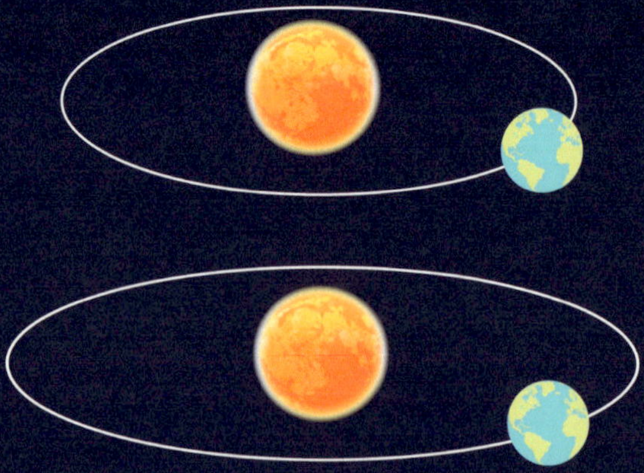

This is one of the biggest pieces in the jigsaw puzzle of climate change and the picture makes no sense if this piece is left out. Why are children taught the CAGW hypothesis, but not the Milankovitch theory? This is akin to teaching children alchemy, whilst ignoring chemistry.

The Sea

The sea is a vast store of fish and shipwrecks and, more important for our purposes, energy and CO_2. Water has some special characteristics that enable it to play a major role in the heat exchange processes of the planet. For example, it takes large amounts of energy to turn ice to water and even more to generate steam. It also retains heat incredibly well, a cup of tea takes some time to cool down to room temperature whereas a plate of fries will be cold in fifteen minutes. Over 70% of the Earth is covered in water and 97 percent of Earth's water is in the oceans, with a mere 2% in the frozen glaciers and ice caps. Less than 1% is fresh water and a tiny fraction is the water vapour in our atmosphere[7].

Perhaps the depths of the sea could replace outer space as the new frontier of exploration because there is so much to learn and oceans have so much impact on the climate and human activities.

The ocean temperature can increase due to solar radiation heating the surface of the water, or heat escaping from the Earth's molten core when undersea volcanoes heat the lowest parts of the oceans. When ocean temperatures rise, CO_2 is released (just like a warm coke bottle) though this is a slow process, following the increasing temperature by 700 to 800 years.

According to alarmist theory this would result in a positive feedback loop and cause runaway warming. That scenario neglects the "smart" negative feedbacks, the Earth's checks and balances and failsafe's that stabilize the system. Warming oceans increase evaporation and cloud cover (the sunshade) thus preventing a chain reaction. This prevents the seas from boiling away over millions of years.

There are numerous positive and negative feedback loops in action within Earth's atmosphere but alarmists highlight the positive feedbacks of CO_2 and ignore or downplay the negative feedbacks. More research is needed but it is abundantly clear from the geological record that the planet is far more likely to get cold thanks to positive feedbacks from cooling rather than experiencing runaway warming, regardless of the source.

Positive and negative feedbacks are confusing if people incorrectly assume that positive feedback means an increase (in temperatures) and negative a decrease. However, positive feedbacks can work in both directions, either warming or cooling.

A negative feedback is like a double electric blanket. The man and wife manage the temperature on their side of the bed keeping it stable and comfortable. Imagine if they accidently swap their controls. Now we have a positive feedback loop. When the man is too warm, he turns the temperature down but that makes the wife colder so she turns the temperature up only it is actually controlling the husbands side so he gets even hotter and turns the temperature down again making the wife even colder! This positive feedback loop eventually spirals out of control into instability.

We have not explored very much of the 72% of the planet hidden by the oceans. We are still learning new things all the time and large tracts of the ocean floor are not yet mapped. We know there are around 1500 volcanoes on the surface but there are an estimated 1,000,000 (a million) undersea volcanoes. A hundred volcanoes have been discovered under Antarctica[8] alone and the largest volcano in the entire Solar System was discovered in 2013, a thousand miles off the East coast of Japan below 2,000 feet of water[9]. It is about the size of the British Isles. Oceanographers have found some extremely strong correlations between undersea volcanic eruptions and increased melting of ice around the ice caps. Other scientists believe there is a strong link between Ice Ages, undersea volcanoes, and sea level changes[10].

According to climate science orthodoxy, melting ice is the fault of a half degree of warming in the air, whereas millions of undersea volcanoes emitting millions of tons of lava at 700 to 1,200 °C (1,292 to 2,192 °F) for days, even months on end have absolutely no effect on sea temperatures nor the natural CO_2 levels.

Like everything else in nature, the oceans have multiple cycles and fluxes that distribute heat and moisture across the world. There are currents and streams such as the Gulf Stream which circulate colder and warmer waters constantly. The various combinations of seas, winds, currents, humidity and land drive climatic events from thunderstorms to hurricanes. There are numerous recurring patterns such as the Pacific Decadal Oscillation (PDO) that occur over periods from days to decades and some that are even longer[11]. Oceanographers only know roughly when these changes occur. Some irregular events that disrupt normal climate patterns are more localised such as El Niño (warm sea surface) and La Niña (cold sea surface) in the Pacific.

The crucial thing to remember is that while the ocean surface heats the air, the air cannot heat the oceans in any significant way due to the massive difference in the heat content of the oceans compared with the atmosphere. How therefore could a tiny amount of CO_2 in the air raise the temperature of 1.3 billion cubic kilometres of water in our oceans and seas? Most of the deep ocean is only a few degrees above freezing: warmer air and water tend to rise, not sink and a colder body (the air) cannot heat a warmer one (the water). Despite these nearly freezing temperatures at the bottom, the seas contain more than 50 times the heat that is in the atmosphere[12] and can store and release this heat over long periods of time. Think about the bath example again. If you run a hot bath in a cold room the air temperature in the bathroom will increase after a while and the bath water will still be warm enough to enjoy. The reverse is not true. A cold bath in a hot room will not be ready for you no matter how long you wait. The oceans and atmosphere work the same way. One cannot simply dismiss these everyday observations, the laws of thermodynamics work the same whether in a cup, bathtub, swimming pool or sea.

And Everything Else...

There are many other phenomenon that influence the climate in complex ways, interacting, counteracting, amplifying and reducing each other. There are a slew of other hypotheses that have yet to be investigated enough to be refuted or substantiated. We have not discussed the volcanoes on the surface which occasionally erupt, releasing CO_2, sulphur and dust that can cause major catastrophes such as the postulated eruptions that have blanketed the Earth in dust so thick that the planet was shrouded in darkness.

There are even galactic cycles that affect climate. Cosmic rays fluctuate depending on the position of the Earth within the galaxy and they influence cloud formation: more clouds have a cooling effect by reflecting more heat back into space. These cycles correlate very strongly with Earth's six Ice Ages.

Ice itself has a complicated role in climate change and once again, the alarmists have it all wrong and back-to-front. Paradoxically, melting glaciers and ice caps are not necessarily a sign of warming; instead this can mean the atmosphere is cooling and vice versa. Cooler temperatures results in less evaporation and cloud cover and more solar energy reaching the planet whilst warmer temperatures mean more evaporation and increased cloud cover with less solar energy. These conditions may not persist owing to the complex feedback loops of the climate.

When we add in other factors such as the Coriolis effect, shifting continents, rising and sinking land masses, biodiversity, ozone, the odd stray comets that strike the Earth, the gravitational influences of the other planets and other yet-to-be-discovered phenomena, it is not credible to suggest that one minor factor – the 400ppm of atmospheric CO_2 – overwhelms and subdues all other influences.

The Greenhouse Effect?

So, what is the role of the so-called Green House Effect (GHE)?

According to the hypothesis, solar radiation, enters Earth's atmosphere and hits the surface. The ground absorbs this radiation, becomes warmer, and re-emits some of the energy back into the atmosphere. The GHGs in the atmosphere (remember; mostly water vapour with CO_2 playing only an exceedingly small part) then re-emit some of that radiation which supposedly warms our atmosphere from -16 °C to the cool +14 °C it is today.

It sounds simple and plausible but there are several issues here.

Firstly, the Earth is not a Greenhouse. No astronaut ever reported breaking through a glass ceiling as they passed through the stratosphere, mesosphere, thermosphere or exosphere, not even the female astronauts. This is not nit-picking because the term itself "Greenhouse Effect" is not appropriate, it is very misleading, and confusing, so much so that it should really be renamed. A garden greenhouse works by allowing heat radiation in and blocking the escape of warm air to the outside.

As a matter of interest, growers often inject additional CO_2 into the glasshouse to achieve three or four times the atmospheric level - around 1500ppm. This does not increase the temperature; it makes up for the less than optimal amount of CO_2 in the air outside and the result is the plants grow much faster and larger.

In the atmosphere some of the re-radiated heat from the earth on the way into space is "trapped" by GHGs but the effect is very small. The Earth's greenhouse is more like a screen door with the screen part being made of extremely thin strands of reflective silk.

Layers of the Atmosphere

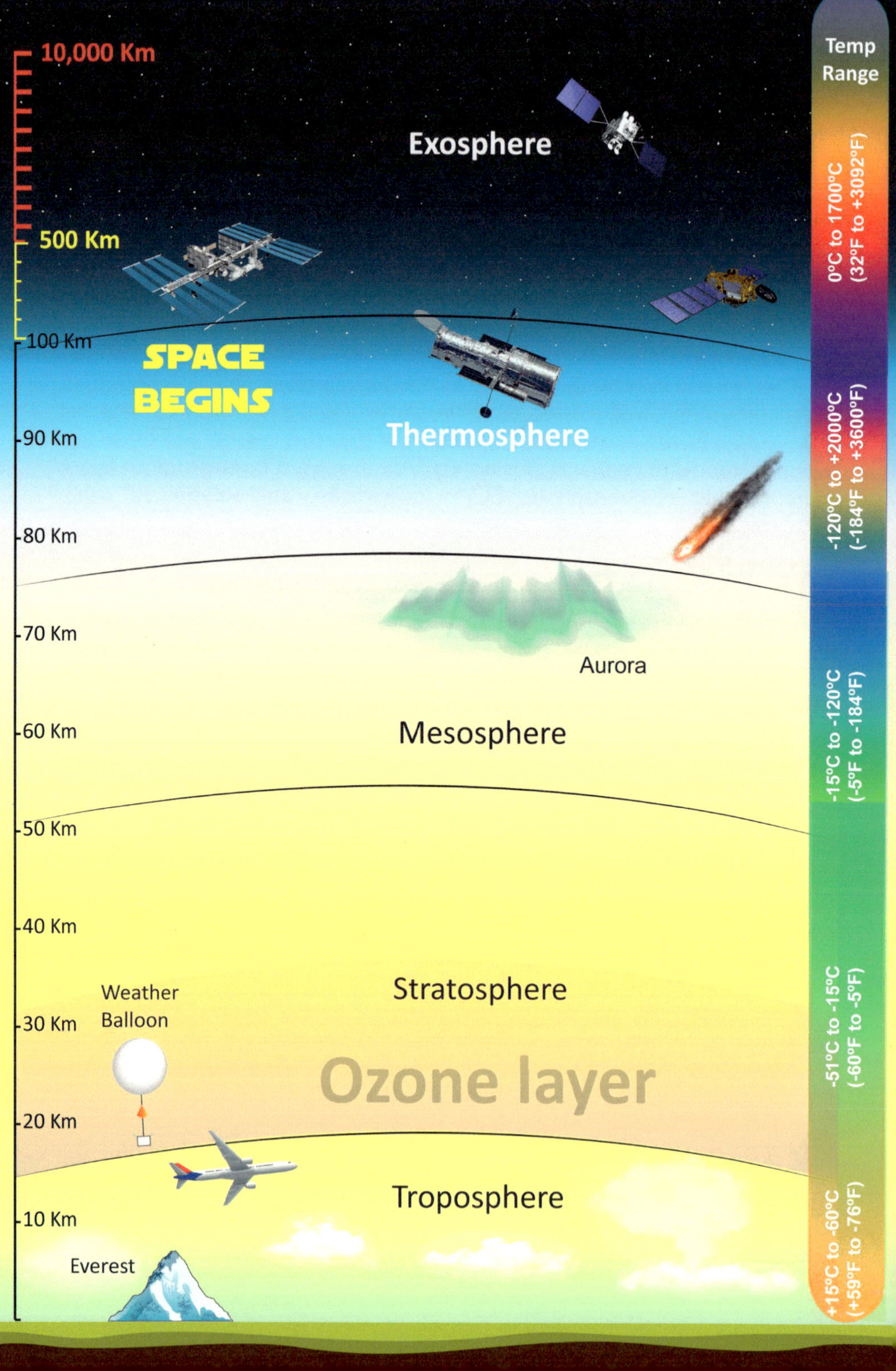

Secondly, the GHE works both ways. When radiation enters the atmosphere the GHGs absorb and re-emit some of the radiation before it reaches the ground. It does this in all directions, including back into space, and more radiation is reflected into space than reaches the Earths' surface.

If you removed all the main Greenhouse gases (water vapour, methane, CO_2, nitrous oxide) then the lower atmosphere (stratosphere) would be much colder but the surface temperature (troposphere) would be much hotter during the day and the opposite at night – basically like the desert. This is because water vapour is also the main mechanism by which the planet cools as well as being a greenhouse gas. Of course, as previously mentioned, without CO_2 no life would exist on Earth anyway.

Warm air rises and as the air pressure decreases it expands and cools condensing into water vapour forming clouds. The kinetic energy of the warm air temperature gets exchanged for the potential energy of its height which eventually falls as rain or snow. Clouds (water vapour) are like sweat; they regulate the temperature in a negative feedback.

The clouds not only reflect radiation from the Sun, they can also re-radiate heat from the planet into space. Paradoxically, the GHE reduces, or slows down the rate of cooling. This is not the same as warming. This may sound confusing, so consider what happens when you set your fridge from the highest setting (very cold) to the lowest setting (mild). The temperature in the fridge will rise but you are not heating the fridge by adding a new source of warming.

Finally, the atmosphere becomes thinner and colder the higher you go. Anyone who has flown a commercial plane will know this by looking at the interactive maps that tell you the speed, location and outside air temperature, that ranges from -40°C to -57°C (-40°F to -70°F). Commercial airlines fly at 9-13km (35,000-42,000 feet) in the stratosphere, just above troposphere, to avoid the weather and turbulence. The thinner air means less fuel is required while there is enough oxygen to keep the engines working.

The GHE postulates that the gases in the freezing cold stratosphere trap some of the heat radiated from the Earth and send it back to earth to warm the ground further. This is a fundamental error because you cannot add temperatures like that. It contradicts the Laws of Thermodynamics and therefore the Laws of Physics. As Chief Engineer Montgomery Scott from *Star Trek* would say "I canna change the laws of physics". Apparently, climate alarmists think they can.

Imagine the result if this back radiation heats the already warm ground and causes it to emit more heat, and some of that is trapped and reflected back by GHGs again? This would be like a ping pong ball forever bouncing between two surfaces and the Earth would have long ago boiled away were this to happen.

At some point an equilibrium is achieved, as you can understand by the simple experiment of putting a blanket on a central heating type radiator. This does not warm the radiator more which then warms the blanket more and so on until the blanket catches fire. The blanket also radiates heat until it reaches equilibrium.

However, the GHE of alarmism depends on the notion of heat warming up heat, as though the temperature that results from mixing the heat in two bodies will be the arithmetic sum of their individual temperatures. This is like pouring a cup of boiling water (100°C) into another cup of boiling water (100°C) and claiming the water is now 200°C. Or turning on one bar of a heater that heats the room from 15°C to 30°C then turning on a second bar. Does the temperature then rise to 45 °C? Or 60 °C with a third bar?

There is an engrossing academic debate, especially amongst physicists and highly specialized scientific nerds over the GHE and whether it warms or cools or is effectively insignificant. The main result of this debate has been to substantially reduce the consensus on the most likely "forcing effect" of increasing atmospheric CO_2 from the unrealistic figure suggested in the early IPCC Assessment Reports.

So, the science is most certainly not settled in this area. It is clear that we cannot view these various warming and cooling mechanisms as standalone systems because there are so many interactions and the net effect is to maintain a remarkably stable temperature over the vast expanse of geological history. The Earth really is a brilliantly designed symbiotic system. The misnamed GHE saves the Earth from warming or cooling too rapidly and enables gravity to create a temperature gradient sufficient to keep the surface warm enough for life whilst water in its three phases (solid, liquid and gas) maintains the distribution and transformation of heat.[13]

Conclusion

We need to be clear about what we know and what we do not know. At the frontier of knowledge there will always be "unsettled science". Contrary to the claim that the science is settled in favour of alarm about warming we do know enough to be confident that CO_2 is not a major factor.

The purpose of this book is to explain as clearly as possible that anyone claiming they know the full story about climate change has not done their homework in the serious scientific literature. The science surrounding climate change, aka global warming, is certainly not settled.

What we do know is that there are many different cycles that interact, overlap and intertwine with each other, and that unpredictable events often interfere with these cycles. Alarmists claim that all of these variations have been taken into account. This is not true because some of these interactions have only recently been discovered, and they are all topics of active investigation with very many unanswered questions. With all these variables there is simply no way that a credible model could be developed to predict with any accuracy the climate in the next few years, or even decades.

Yet somehow, out of this intricate web of interwoven, interacting threads we are led to believe, that the single gossamer-thin strand of CO_2 is the principal driver of climate. Even if one buys into the GHE hypothesis, it should be obvious that the contribution of manmade CO_2 is so small as to be virtually undetectable.

Compared with other planets Earth's climate is relatively stable. There have been Ice Ages and warm periods of relative cold or warmth in the past, but the first chart in the *Some Like It Hot* chapter shows that there was no runaway warming regardless of the level of CO_2. Fluctuations in CO_2 without runaway warming are entirely natural phenomena that have been occurring for billions of years and will continue for billions more. The checks and balances within both long-term cycles and the everyday climate ensure the planet remains habitable for life.

Chapter VII
PULP FICTION
Fake Science, fake news and the Politics of the IPCC

"The prospect of domination of the nation's scholars by Federal employment, project allocations, and the power of money is ever present – and is gravely to be regarded. Yet, in holding scientific research and discovery in respect, as we should, we must also be alert to the equal and opposite danger that public policy could itself become the captive of a scientific-technological elite."

PRESIDENT DWIGHT EISENHOWER

The IPCC is probably the most influential driver of alarm because it is widely regarded as a reputable scientific body. In reality it is a political organization that was established in 1988 to provide the appearance of scientific support for the long list of agencies and activities that the U.N. established to promote sustainable development and actions to limit perceived warming. One might expect that the members of the IPCC are actually working scientists but the membership consists of two political appointees from each of the current 193 member nations in the U.N. Their identities and their qualifications are not readily accessible or widely known and they are individually unaccountable due to the collective nature of the Panel's decisions. The prime movers in addition to these political appointees are the activists in the U.N. who organize the proceedings of the Panel and recruit suitable scientists to support the agenda. These activists are the full-time U.N. IPCC staffers who are the gatekeepers, monitors and directors of the entire proceedings.

The Panel produces major Assessment Reports (AR) at regular intervals. These are sometimes called the "Climate Bible", the gold standard for climate science, because they are supposed to be based on the pooled wisdom of thousands of climate scientists. The first and shortest edition of the IPCC Assessment Reports appeared in 1990. Its findings were tentative and speculative yet by June 1992 they had become imperatives requiring urgent action. In the climate of opinion promoted by environmental activists around the world, the U.N. convinced a majority of the world's governments to sign the *United Nations Framework Convention on Climate Change* that declared greenhouse gases to be the latest and greatest menace to the planet.

The Assessment Reports appear in three volumes prepared by three Working Groups (WG). WGI covers the physical science, WGII the impacts and adaptation and WGIII the mitigation of climate change. Note here that the assumption is already made that man-made global warming is the driver of climate, has measurable impacts and is dangerous and must be mitigated. The IPCC is not there to assess whether there is global warming or whether it is anthropogenic, they start with the conclusion and work from there. That is not science.

Since the Assessment Reports are over 2000 pages long a short summary volume is produced that provides policy advice based on alarming projections into the future. This summary is devoured by politicians, green activists and the media. The Assessment Report is supposed to condense the major findings in a thick volume of supporting papers but this has next to no circulation and as far as the politicians and the public are concerned it might as well not exist.

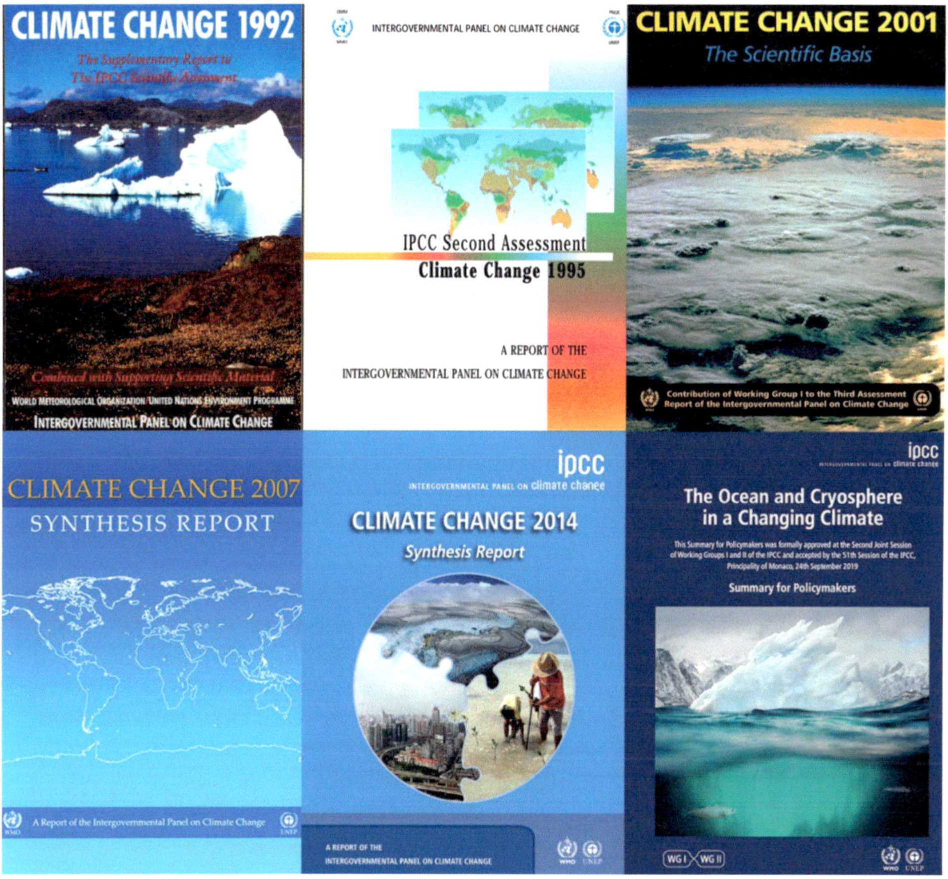

Some of the voluminous IPCC Reports that have been produced over the past three decades.

The summary statements are crafted to provide ammunition for radical activists and headlines for stories, mostly written by scientifically illiterate journalists, to frighten politicians and the general public. The text of the all-important summary is crafted in a line-by-line approval process in gruelling days of mass meetings where the members of the Panel labour under the supervision of U.N. officials and monitors.

Dr. David Legates, American climatologist and professor of geography at the University of Delaware describes the summary document thus,

> There's a science document which is really written by scientists for scientists. There's also a summary for policymakers, it is put together by policymakers and in many cases they go back to the scientists and say 'can you change the science document to match our summary. We want to beef this up, we want to make it look worse'. That's not the way science is done.[1]

The short summary volume must be distinguished from the bulky volume of substantial papers that are supposed to provide the foundation for the summary and recommendations. The Panel maintains some veneer of scientific respectability due to the higher quality of many of the background papers where the control of the final product is not as strict as the dictation of the summary report. However, the Panel itself is not immune from criticism or error. One of the red flags regarding the scientific credibility of the Panel is the list of distinguished climate scientists who have dissociated themselves from it, have been side-lined or have been involved in the process but were frustrated in their efforts to contribute effectively.

After the publication of the Second AR in 1995 Professor Frederick Seitz, former president of the *National Academy of Sciences*, said "I have never witnessed a more disturbing corruption of the peer-review process than the events which led up to this IPCC report." Fifteen passages had been deleted from the document that had been approved by all 28 contributing authors. What was so controversial? Well every scientist had expressed considerable doubt about man-made global warming. Two passages were as follows,

> None of the studies cited above has shown clear evidence that we can attribute the observed [climate] changes to the specific cause of increase in greenhouse gases.
> No study to date has positively attributed all or part [of the climate change observed] to [man-made] causes.[2]

Dr. Roger J Pielke resigned in 1995 during the release of the Second AR. Pielke had recommended texts and papers that the IPCC simply ignored and he was appalled at the use of computer model predictions that he had pointed out were unproven and untrustworthy yet which the IPCC claimed were certain.[3] Pielke was later proven correct as the computer models failed over and over to make any accurate predictions. Interestingly, the use of and reporting on climate models follows a very similar process to the IPCC selection of papers and authors. The collated evaluations are called the CMIP (Coupled Model Intercomparison Project) and a report is released (that corresponds roughly with the release of the IPCC main Assessment Report) based on the models that they have accepted as part of CMIP. Not surprisingly some models are not accepted to the CMIP project, presumably because they do not predict catastrophic warming.

Atmospheric physicist Dr. Richard Lindzen, resigned from the IPCC in 2001 because the Summary Statements did not represent what was reported in the larger volume of technical papers.[4] Professor Richard Tol resigned in 2014. He is a leading economist who extensively modelled the likely impact of warming. In his notice of resignation, he wrote that the Summary Statement was the result of encouraging over-estimation of hazards due to self-selection of authors and referees and that the report;

Omitted improved irrigation and crop yields.
Emphasized heat stress but downplayed reduced cold stress.
Warned about mass climate migrations without any solid evidence.
Overestimated the consequences of climate change.
Could have been written by the Horsemen of the Apocalypse.[5]

There have been a number of other resignations, complaints and attempts to reform the IPCC because of the way it has misrepresented, omitted, obfuscated and exaggerated. Yet, it is still held up as an august institution and was even awarded the 2007 Nobel Peace Prize just as controversy over its workings started to boil over.

The Laframboise Study[6]

Canadian investigative journalist Donna Laframboise has investigated many aspects of the organization and management of the IPCC procedures and practices to explain why so much of their output lacks scientific credibility.

She cited widespread claims that the input to the IPCC Assessment reports met the highest standards of the peer reviewed literature. We are told that the science has been settled based on "an immense edifice of painstaking studies published in the world's leading peer-reviewed journals...vetted and documented in excruciating detail..."[7] A climate modeler claims that the input to the reports "has been scrutinized to the highest level possible" and "a core principle of the IPCC is that only peer-reviewed literature is cited"[8].

Laframboise described a collaborative project involving a worldwide team of unpaid "citizen auditors" who checked all the references cited in the 44 chapters of the 2007 report to find how many were peer-reviewed and how many came from the "grey" literature. Out of 18,531 references 5,587 (one third) were not peer reviewed. Among the sources used to support IPCC recommendations were newspapers and magazine articles, unpublished theses, *Greenpeace* and *World Wildlife Fund* documents and even press releases.

She wrote,

> While Chairman Pachauri had declared an Indian government discussion paper fit only for the rubbish bin, we found that the Climate Bible cites dozens of discussion papers. In one case the document relied on by the IPCC was clearly labelled as 'version one' of a draft.[9]

After the release of Laframboise's findings the IPCC Chairman wrote a promotional piece referring to "approximately 18,000 peer-reviewed publications", with the concession that "a limited amount of grey literature" was used "in cases where peer-reviewed literature was unavailable". He referred to 18,000 peer reviewed publications when he should have said 18,000 minus 5,500. Is a third "a limited amount"? And how did it happen that of all the thousands of climate scientists in the world who, according to the oft repeated claim, 97% of which agree with each other, left the IPCC short of peer-reviewed material for the report so they had to resort to material from green activists?

There is supposedly a house rule that limits the quantities of non-peer reviewed material and another "rule" that such items would be clearly identified in the list of references. One of Laframboise's volunteers found that just six of almost 6,000 grey items were flagged. She learned from some people close to the Panel that more transparency was suggested, perhaps using a colour code to flag grey material. After some internal discussion there was a policy statement that peer-reviewed material should have priority and there was no more mention of flagging grey literature.

Laframboise explained the various categories of writers involved in writing the papers in the major volume. These have more merit than the summary but the final writing is in the hands of the "right" people. Coordinating lead authors (usually two) oversee each chapter. Lead authors, ranging in number from a handful to dozens, do the bulk of the writing. Contributing authors provide material on specific topics to be cited or re-worked. They usually do not attend the meetings of the other authors and so they have no say in the way their work is used and edited.

The coordinating lead authors are the key figures, followed by the lead authors. Some of these are very unlikely contributors in terms of experience, qualifications and background. Richard Klein achieved the status of a top world expert and coordinating lead author long before he completed his doctorate[10] and not long after working for Greenpeace. And Lisa Alexander was still completing her PhD at Monash after she had been a contributing author[11] to the 2001 Bible and a lead author in 2007.

Along with the deception regarding the amount of peer-reviewed literature in the references these discoveries about the major contributors should have drawn some reportage but there is a dearth of scientifically literate, responsible and professional science journalists.

Bending the rules

Some scientists have highlighted the academic bullying and malfeasance practised by a few of the IPCC Lead Authors. In 2005, leading hurricane expert Chris Landsea (a great name for a hurricane expert!) resigned from the IPCC after Dr. Kevin Trenberth used his position as a lead author to undermine his research and present a completely different story to the media. Landsea and others found no long-term trend in cyclones or hurricanes increasing and even their worst-case scenario for any future warming was a slight increase in the amount of rain falling during a hurricane. Trenberth gave a totally opposite report to the media, who of course accepted unquestioningly, claiming hurricanes were increasing in frequency and intensity and it was caused by man-made global warming.

Other tales of fabricated graphs, of submitted papers being described as "the worst paper I have ever read in a major scientific journal"[12] and the insertion of uncorroborated claims such as the Himalayan glaciers melting[13] have further called into question the integrity of the organisation. The shortcomings and failures by the IPCC led American physicist Dr. Fred Singer to set up a parallel organisation called The Nongovernmental International Panel on Climate Change (NIPCC)[14] which publishes its own hefty report written by teams of scientists and economists across the globe. Amongst the NIPCC lead authors is Patrick Moore, one of the founders of Greenpeace. The NIPCC critiques the IPCC reports in detail and debunks many of their more alarmist claims. That a group of scientists were so concerned about the flaws and errors emanating from the IPCC that they voluntarily write and develop their own report, unpaid, speaks volumes to how badly the IPCC has failed.

The Dangerous Liaison of Science and Politics

Is it science that has hijacked politics or politics that has hijacked science? The funding and government influence over science and technology has grown massively since the end of World War Two and whilst President Eisenhower's farewell address is remembered more for the coining of the term 'military industrial complex' and his warnings against its growing influence, he also made the remarkably prescient warning about the politicization of science that begins this chapter.

The Military Industrial Complex is recognised widely today and the narrative goes that defence companies have huge influence in Washington because war is profitable for them. Why then does this not apply to the Climate Change Industrial Complex, especially given the huge sums of money that are now on the table for all manner of individuals, organisations and corporations?

Both things Eisenhower feared have come about.

Scholars and academics have been lured by money & employment opportunities and in turn public policy is being dominated by an elite clique of scientists. Many scientists who prefer evidence to money have quit the IPCC in frustration and the few that remain have their conclusions watered down, ignored or buried altogether. What is left is a small cabal of alarmist scientists who push exaggerated claims based on faulty computer models and logical fallacies. If the IPCC did not make claims of catastrophic global warming then there would be no need for it to exist. The scientists and bureaucrats would be out of their well-paid jobs, out of the media limelight, and the five-star luxury conferences in exotic locations would be no more.

Thus, a self-perpetuating machine controlled by politicians and a gaggle of doom merchants and powered by money hides behind a veneer of scientific authority. And an unelected transnational organisation holds sway over vast areas of the economies and policies of every country on Earth.

Chapter VIII
LA-LA LAND
The Computer Models

"This is the world that you know. The world as it was at the end of the twentieth century. It exists now only as part of a neural-interactive simulation that we call The Matrix.
You've been living in a dream world, Neo."

MORPHEUS, THE MATRIX (1999)

It is amazing how often the spectre of artificial intelligence (AI), advanced computing and machines taking over the world crops up in the media and popular culture these days. If the global warming models are a benchmark of our computational and programming abilities then we have absolutely no need to fear super-intelligent cyborgs, silicon enslavement or AI instigated nuclear war. The climate computer models are the entire foundation of all the predictions, exaggerated claims and apocalyptic scaremongering emanating from the alarmists and amplified by the media. Over one hundred different computer models have all seemingly predicted similar outcomes many years into the future and all, so far, have proven to be wildly inaccurate.

There is nothing special, mystical or magical about a computer model or even a computer itself. Whilst we can do amazing things using computers, from online shopping to making fully realistic virtual worlds, they are, at the end of the day, just very powerful calculators that require instructions and rules to operate. These instructions and rules are called programs and a climate model is merely that - a set of rules and instructions contained in a program. At the end of the day the computer is merely carrying out the commands and directives of its human masters. And since programs are written by humans they are susceptible to errors and biases from the outset. It is a grave mistake to believe computers are infallible.

Models can be extraordinarily complex. Children who build physical models such as ships, planes or spaceships whether it be in wood, molded plastic or Lego bricks understand this. One misplaced piece can cause the whole model to turn out wrong or fall to pieces. And a 1500-piece Lego Millennium Falcon is many orders of magnitude easier to build than a global climate model. Furthermore, Lego comes with instruction booklets that show how all the pieces fit together and in what order to assemble them. Try deviating just slightly from the instructions, skip a few pages or use the wrong piece and see how it turns out. The instruction book is crucial.

The problem with climate models is that we do not have an instruction book. We do not fully understand how the climate works or how the myriad of inputs and variables fit together and scientists will readily admit this. Weather and climate are chaotic and the term "chaos theory" is often used to describe these processes. Chaos is a Greek word meaning "vast chasm" or "void" and today means something that is so disorderly or random that it cannot be understood or predicted. And here is the critical point; You cannot build a reliable model of something you do not understand.

So, if weather and climate cannot be properly understood or predicted then how can you build a predictive model and claim it is accurate? Well actually the IPCC themselves fully admit that the climate models are unreliable and should not be trusted:

> In climate research & modelling, we should recognise that we are dealing with a coupled nonlinear chaotic system, and therefore that the long-term prediction of future climate states is not possible.[1]

Yet they and then the media and the gullible public fall into the same trap, proclaiming these predictive models as the gospel truth.

This is the equivalent of reading the blurb at the beginning or end of movies; "the characters and events in this movie are fictional", ignoring it and then insisting that the movie is real. It is no wonder that the lines between reality and fiction are so often blurred and how someone can watch a fictional movie and believe it will become reality one day. *Waterworld, The Day After Tomorrow* and *An Inconvenient Truth* are good examples.

If we only have a partial knowledge of climate systems we cannot make accurate computer models that rely on knowing everything about the climate. This would be like knowing everything about the properties of wood, some things about bricks and mortar, a little knowledge about glass and the properties of steel are a complete mystery and yet we decide to build a skyscraper. Would you be confident that the building is safe? Would you risk taking an elevator to the top?

The NASA space shuttles were astonishingly complex designs with every piece, all materials and moving parts seemingly known to engineers yet there were still two fatal accidents, Challenger (1986) and Columbia (2003). Would they have launched missions if they had to guess the insulation properties of the tiles or assume the mixtures of liquid oxygen and hydrogen without knowing their properties?

That is pretty much where we are at with climate models; guesswork and assumptions based on vague correlations, unproven hypotheses and an arrogance that we know far more than we actually do.

So exactly how wrong have the climate models been? The chart below compares what the models have predicted over the past three decades versus the reality.

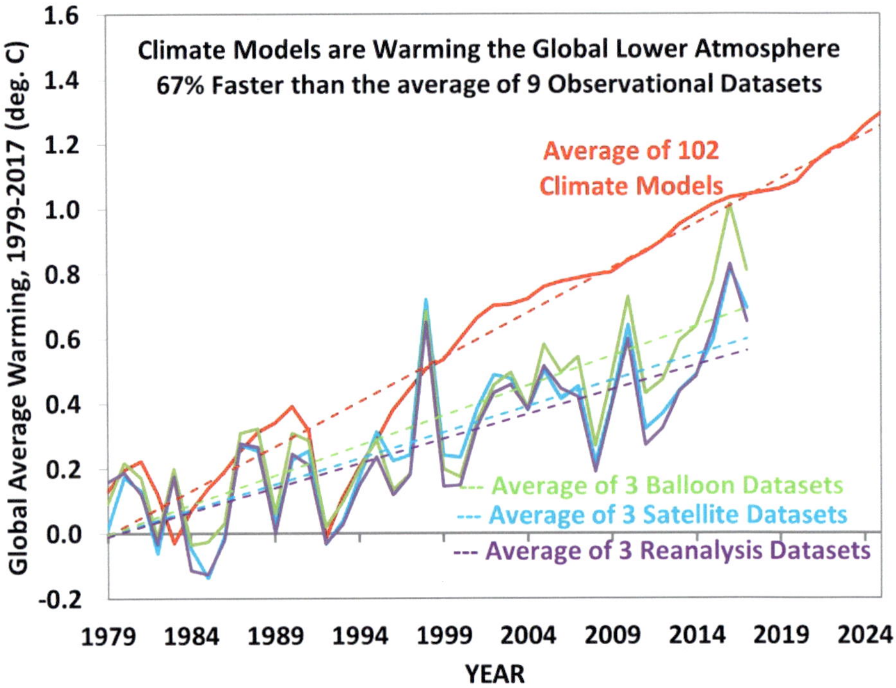

The red dotted line shows the average of all predictions made by the global warming computer models. The other dotted lines are the average actual temperature as measured by satellites and weather balloons. Virtual reality and reality are diverging to an increasingly greater degree.[2]

It is interesting to observe that the earlier models predicted warming and cooling but since the early 1990s they have only predicted warming. This was around the time that funding for scientists to find global warming began to increase exponentially. Is this correlation or causation? What do you think the models would predict if funding was directed to find cooling instead of warming?

The prediction models are clearly running way too hot. Some have speculated that the models do not model clouds or the sun correctly and others have pointed out glaring omissions of crucial factors such as submarine volcanoes. Despite being funded by taxpayer's money the code and data used in the models is kept largely secret. This makes a mockery of science and the scientific method. No teacher would accept a pupil's mathematics or science homework which just showed the answers without the working out and nobody would have accepted Newton's assertion that objects accelerate at $9.8 m/s^2$ unless he had submitted his equations and data for other scientists to check and confirm.

In June 2019 one whistle-blower blew the lid off the computer models. Dr. Mototaka Nakamura published *Confessions of a climate scientist: The Global Warming Hypothesis is an Unproven Hypothesis*[3] which exposed how fatally flawed computer modelling of the climate is. Nakamura, with decades of experience in computer modelling of clouds and oceans, concluded that the climate models the IPCC relies on make very simplistic assumptions, omit key factors, do not have reliable data and could not possibly perform all the required calculations owing to the limitations of computing. "Toys" and "Mickey Mouse mockeries" are how Dr. Nakamura describes the models and given how the models continually require "tuning" of parameters this is an accurate assessment. This is somewhat like computer games such as *Elder Scrolls*, *Grand Theft Auto* and *Minecraft* which allow users to tune the climate with the press of a few buttons. These massive open world games simulate real-time weather and climate that can affect the physics and outcomes of the game. Are we all being deceived by computer simulations? According to Dr. Nakamura and others the answer is yes.

Players can cycle through a range of weather condition in games such as Grand Theft Auto and Minecraft. The weather patterns affect the game.[4]

The most astonishing simplification the computer models make is to assume the output of the sun is constant. This is patently absurd. The Earth receives vastly different amounts of radiation depending on the output of the sun, the cloud cover, the distance from the sun and the rotation, tilt and wobble of the Earth. Apparently, none of this is important enough for the computer models but a fraction of a percent of CO_2 is a critical variable. Solar irradiance is very strongly correlated to the temperature of the Earth, much more so than CO_2 and historically colder periods have happened when the Sun has taken a nap. We entered just such a cycle in 2020 which may last for several decades and which will bring colder and prolonged winters to the Northern hemisphere, according to some astrophysicists.

The sun is clearly the main driver of the Earth's climate but there are also a myriad of other factors. Tides, water vapour, the oceans, wind, rain, ice caps, lots of different gases, gravity, volcanoes, plants, trees, grass, animals, people, hydrocarbons, the moon, the orbit, axis and spin of the Earth, cosmic rays – a seemingly endless list of things that change at varying speeds in multiple locations across the globe that all interact and influence each other in countless ways. Even if we understood and could predict the behaviour of the sun and all the other contributing factors accurately there is not a computer in existence that could possibly sort, evaluate and process the seemingly infinite number of calculations and outcomes that could possibly be produced. Theoretically only a very large quantum computer could do this and it would still take an unfathomable amount of time and energy. There is not even agreement on all the possible outcomes of a chess game or the Chinese game of *Go*. No computer program exists that can accurately calculate all those outcomes and these are games that are many, many orders of magnitude less complex than the climate. And while we are 100% certain about the rules and parameters of a game of chess or *Go*, there are plenty of factors relating to climate of which we know very little or nothing at all.

Prediction and optimisation models are extraordinarily complex and exceedingly difficult to build for even short-term predictions of comparatively simple processes. The tiniest of inaccuracies or one missing variable of thousands could skew results wildly, particularly over long time frames. An inaccuracy of just one hundredth of one degree (0.001) per day would produce warming (or cooling) of 8° in a climate prediction twenty years' off. Bad models also produce bizarre phenomenon and warped scenarios. It is not surprising therefore that the climate models predicted a tropical mid-troposphere "hot spot" in model after model that is not evident in observed reality.

The model aside there is also the question of data. Predictive computer models require historical data to tune and optimise the model, the more data the better. The data, of course, needs to be accurate, clean and uniform. In the case of climate we have virtually no data at all. We only have accurate temperature measurements for just over a century, and mere decades in the case of sea levels, wind and rainfall patterns, cosmic rays and the numerous other factors that influence the climate. The data is also sparse, having only been collected in a limited number of locations across the globe. Data is currently used from just ten surface temperature stations in the southern hemisphere. Seven of these are in Australia and four of them show an unadjusted cooling trend since measurements began.[5] These pitifully small and limited data sets collected over, at most, a century are a grain of sand in the desert of the Earth's history. Further, the meagre data we have collected is being tampered with, erased and is untrustworthy[6]. "The global surface mean temperature-change data no longer have any scientific value and are nothing except a propaganda tool to the public," writes Dr. Nakamura.

Some critics of the models proclaim "garbage in, garbage out" but even if we had all the most accurate data, stretching back millennia, covering every climate variable, the models would still fail because the model itself is incorrect. Bad data into a bad model will mean "garbage out" and so will perfect data into a bad model. And despite the IPCC itself warning of the unreliability of climate models, despite the problems with the data, and despite the other gargantuan hurdles yet to be overcome with computer modelling, we are asked to trust these predictions. And organisations that we are expected to trust fully support these flawed models. For example, The Australian Academy of Science's 2015 *Science of Climate Change* booklet claims "models provide compelling evidence that recent changes are due to increased greenhouse gas concentrations in the atmosphere."[7]

Surely with over one hundred of these computer models at least a few might have been able to make an accurate prediction? Well in a word, no. It is possible that the climate change community thought that by using one hundred plus models it would somehow bolster their argument. "Over one hundred computer models predict disaster" they would claim, yet this appears to have backfired spectacularly since it has just proven they are wrong one hundred times over. The apparent "97% of climate scientists" are basing their claims on computer models that are 100% wrong. Statistically speaking, you would have a better chance of asking one hundred *Victoria's Secret* models to come up with a prediction of what the climate will do in ten or twenty years as the answers would be so random that by sheer accident a couple of them would probably stumble on something close to reality.

Victoria's Secret models. Smarter and more accurate than global warming models. [8]

To conclude, there are three basic requirements for a reliable climate model: the model itself, accurate and comprehensive data sets and an appropriate supercomputer. The global warming industry has none of these, it is not even close in any of the three pre-requisites but we are continually told "the science is settled". The model cannot be built because we do not fully understand how climate works, something all climate scientists admit. The accurate data sets do not exist and what little data we have is being fiddled with and the supercomputer required to crunch the numbers does not exist and for all intents and purposes would be impossible to build let alone run. In fact, these climate models are an affront to science, they should never have been invented. Splicing the genes of a man, a bear and a pig together to create a hybrid man-bear-pig monster is more ethical and more conceivable than the climate models. *Victoria's Secrets* models would be better predictors of climate than bad computer models and they would not triple our power bills and cripple the economy.

The climate models are an artificial computer world that generate an apocalyptic future to instil fear into people so that they will turn to Big Government for protection. In *The Matrix* Morpheus "red pills" Neo by waking him up to the fact that he is living in an artificially constructed dream world in order to keep him as a slave to power the machines that control him. He even warns Neo that there are people hopelessly dependent on the artificial world for survival and they will fight to keep it alive.

Some movies, it seems, are not without a sense of irony.

Chapter IX
RAGING BULL
The Failed Prophecies

"The whole aim of practical politics is to keep the populace alerted (and hence clamorous to be led to safety) by menacing it with an endless series of hobgoblins, all of them imaginary."

HL MENCKEN

There are so many examples of failed prophecies of doom that we could publish a separate book on them. Yet climate alarmism continues undeterred despite a long record of failed predictions and prophecies. In fact, it is getting worse; the alarmists have learnt nothing and doubled, even tripled down on their fearmongering as if reality did not exist. These false alarms could be regarded as a laughing matter but the results are not amusing in the light of the emotional, psychological, financial, societal and environmental damage that is being done as a result of their influence. They demonstrate the ongoing power of the various "end of the world" obsessions that have spawned alarmist cults over-and-over again throughout history.

Over the last century or so there have been at least four episodes of scaremongering over warming and several over global cooling. Two of these episodes came from the same people, predicting cooling first, then warming. James Hansen and Stephen Schneider led the chorus of alarm about irreversible cooling in the 1970's but later changed horses and mounted the more durable global warming steed. Schneider's 1976 book *The Genesis Strategy* detailed how the globe would get colder, cause food shortages and famines and lead to the deaths of billions. According to the *New York Times Review* the book was "reflecting the consensus of the climatological community"[1]. Sound familiar? Moreover, the book detailed the Roman and Medieval warming periods and the positive impacts on society at the time. These are the periods that were later erased from the past by alarmists attempting to rewrite the historical climate record.

Serial Offenders

Al Gore and Prince Charles are serial offenders in making absurd prophecies.

Gore claimed the Arctic would be ice free during the Summer by 2016. It wasn't and is not likely to be for another million years - maybe. The Arctic continues its natural cyclical melting and freezing that happens during Ice Ages and in fact Arctic ice volume is at its second highest level in over 10,000 years[2]. Gore was backed up by similarly absurd predictions from scientists or individuals at NASA[3], Cambridge University[4], the Department of Oceanography of the U.S. Navy[5] and the National Snow and Ice Data Centre[6]. This is a common tactic – one person at an institution says it but the media reports it as "Cambridge University", as if the entire institution is in agreement.

Images of polar bears on precarious ice floes were littered throughout Gore's movie. Gore has been joined by the World Conservation Union[7] and as recently as 2017 The U.S. Fish and Wildlife Service[8] in claiming polar bears were in danger of going extinct due to global greenhouse gas emissions.

Polar bears are now at their highest numbers in decades.[9]

In March 2009 Prince Charles believed that we only had 100 months for desperate remedial action to avert disaster. In July that was down to 96 months. Sometime later it was down to 88 months and in 2014 he complained "We are running out of time. How many times have I found myself saying this over recent years?" In 2015 he relaxed the timetable to allow "just 35 years" to avert catastrophic climate change. In 2017 in Malta, he pressed governments to acknowledge the "huge elephant in the room of accelerating climate change". In another statement he referred to climate change as "the wolf at the door", the greatest threat to the planet, so action was required more urgently than ever.

The "last chance" tactic is par for the course in climate alarmism. We researched and counted at least 12 official "last chances"[10] in the past 20 years and that was without even looking very hard. The latest claim is now "twelve years", according to a former New York bartender and a Swedish teenager who missed a lot of school.

Seven Hells Dispelled

Just to illustrate how terrible scientists can be at making predictions, in 1968 the Stanford biologist Paul Ehrlich and John Holdren (later Barrack Obama's science advisor) published *The Population Bomb* and announced;

> In the 1970s hundreds of millions of people will starve to death in spite of any crash programs embarked upon now," and "by the year 2000 the United Kingdom will be simply a small group of impoverished islands, inhabited by some 70 million hungry people.[11]

Ehrlich and Holdren were not the only academics making such wild predictions. Peter Gunter, a professor at North Texas State University, in a 1970 issue of *The Living Wilderness* wrote:

> Demographers agree almost unanimously on the following grim timetable: by 1975 widespread famines will begin in India; these will spread by 1990 to include all of India, Pakistan, China and the Near East, Africa. By the year 2000, or conceivably sooner, South and Central America will exist under famine conditions... By the year 2000, thirty years from now, the entire world, with the exception of Western Europe, North America, and Australia, will be in famine.[12]

In describing *Seven Big Failed Environmentalist Predictions* author Robert Tracinski noted in 2015:

> It has been 45 years now since the first Earth Day...Harvard biologist George Wald told us, "Civilization will end within 15 or 30 years unless immediate action is taken".[13]

The other six failed predictions were Global Cooling, Overpopulation, Mass Starvation, Resource Depletion, Mass Extinction and Renewable Energy.

In the five decades since making these claims, the world population more than doubled and the proportion of chronically undernourished people dropped from a third to a little over 10%. The Green Revolution (also known as the Third Agricultural Revolution) sparked by Agricultural Scientists transformed the production of food in the developing world and reforms in the direction of private property rights and free-markets lifted a billion Chinese people out of poverty in less than two decades. Similar reforms have seen India transform itself rapidly after it rejected the Socialist economic model. Most Latin American countries, which were supposed to be starving fifteen years ago, are now net exporters of grain, fruit and meat. The exception is Venezuela. Once the richest country in the world, Venezuela embraced Socialism at the turn of the 21st Century and now suffers from all the usual effects seen in countries that adopt Socialism: hunger, inflation, poverty, blackouts and political oppression.

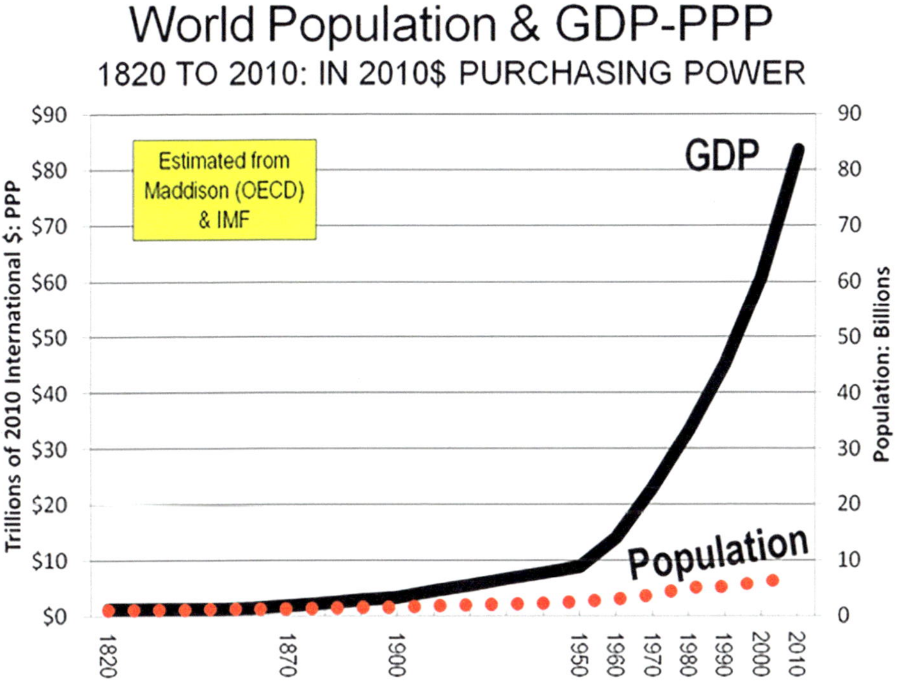

[1]The chart shows the remarkable growth in gross domestic product (GDP) since the start of the industrial revolution vs population growth. People are much wealthier today than at any time in history and the predictions of death and famine never came close to reality.

Going the way of the dinosaurs

At the first Earth Day in 1970, its political sponsor, Democrat Senator Gaylord Nelson, warned:

> Dr. S. Dillon Ripley, secretary of the Smithsonian Institute, believes that in 25 years, somewhere between 75 and 80 percent of all the species of living animals will be extinct.

To put that in perspective, a 75% to 80% mass extinction is on the level of the cataclysm that killed the dinosaurs 65 million years ago, possibly caused by the environmental catastrophe of a six-mile-wide meteor crashing into the Earth and cloaking it in an enormous cloud of ash and dust. Obviously, nothing remotely like that happened between 1970 and 1995.

There are still those who are peddling the notion that we are in the process of the Sixth Great Extinction. So, it did not happen back when they first said it would, but it is going to happen soon, any day now...any day...

Some cold water was just thrown on this theory from an unexpected source: Stewart Brand, founder of the *Whole Earth Catalog*, writes:

> Many now assume that we are in the midst of a human-caused 'Sixth Mass Extinction' to rival the one that killed off the dinosaurs 66 million years ago. But we are not. The five historic mass extinctions eliminated 70 per cent or more of all species in a relatively short time. That is not going on now.[14]

He points out that breathless statistics about the number of species being lost is largely driven by the extinction of already rare species that evolved in isolation and in small numbers on oceanic islands, and which did not survive the contact with the outside world and invasive species:

> The island conservationist Josh Donlan estimates that islands, which are just 3 per cent of the Earth's surface, have been the site of 95 per cent of all bird extinctions since 1600, 90 per cent of reptile extinctions, and 60 per cent of mammal extinctions. Those are horrifying numbers, but the losses are extremely local. They have no effect on the biodiversity and ecological health of the continents and oceans that make up 97 per cent of the Earth.[15]

Moreover, much of this extinction has already occurred since most of the world's islands have already been explored and settled.

Even more failed predictions

Many of the predictions made several decades ago regarding man-made global warming are now shown to be wildly inaccurate. Under the headline *Scientists Say Earth's Warming Could Set Off Wide Disruptions* a New York Times piece in 1995[16] stated "At the most likely rate of rise, some experts say, most of the beaches on the East Coast of the United States would be gone in 25 years".

How is that looking 25 years later?

In 2005, the United Nations Environment Programme predicted that the world would have at least 50 million "climate refugees" by 2010. There have been waves of refugees hitting Europe and America in recent years but none of them are "climate refugees".

A blizzard of claims regarding snow piled up as it was claimed by scientists that the skiing industry would die out[17] and that snowfall will become "a very rare and exciting event". This claim was made by Dr. David Viner in 2000 to which he added "Children just aren't going to know what snow is."[18]

Viner is a senior research scientist at the climatic research unit (CRU) of the University of East Anglia (UEA), the University that became embroiled in the "Climategate" emails scandal of 2009.

UEA
@uniofeastanglia

⚠ Please be aware that due to the extreme weather conditions teaching is cancelled today, Friday 2nd March.

There are serious travel problems across the region and Norfolk Police are advising people not to drive, especially where there is drifting snow.

7:47 PM · Mar 2, 2018 from Norwich, England · Twitter for iPhone

Children born when Viner made his claims in 2000 could well have been the same children who had their classes cancelled at his University in 2018 owing to heavy snowfall. [19]

As record snowfalls hit Europe in 2012 & 2013[20] the alarmists tried to backpedal on their square-wheeled bicycle. In 2014 Dr. John Holdren, co-author of *The Population Bomb* was now director of the *Office of Science and Technology Policy* for the Obama administration and claimed that

> "a growing body of evidence suggests that the kind of extreme cold being experienced by much of the United States as we speak is a pattern we can expect to see with increasing frequency, as global warming continues". [21]

Yes, that is correct. Warming is now making it colder.

Australia has had its fair share of climate clowns. In 2007 Robyn Williams, presenter on the publicly funded ABC *Science Show* claimed the seas could rise up to 100 metres in the next century[22]. In the same year Tim Flannery claimed Brisbane and Sydney could run out of water, that the dams would never fill again and that desalination plants were desperately needed.

Just three years later Wivenhoe dam in Queensland was at 191% of its' normal water supply storage capacity which resulted in the Brisbane floods of 2010-2011 which killed 33, caused $2.4 billion in damage and triggered an estimated $30 billion hit to Australian GDP.[23]

Just as John Holdren was employed by the Obama administration Flannery rather than paying a professional price for his failed predictions, ended up as a highly paid advisor to the Rudd-Gillard Labor Government and was even awarded *Australian of the Year.*

Extreme Weather Events: Fires, Storms and Floods.

"Extreme weather events" are major news items these days as though such things never happened before.

The wild claims of more storms[24], more fires and more hurricanes is yet more fake news. In fact, strong hurricanes decreased and the IPCC AR5 stated:

> Current datasets indicate no significant observed trends in global tropical cyclone frequency over the past century … No robust trends in annual numbers of tropical storms, hurricanes and major hurricanes counts have been identified over the past 100 years in the North Atlantic basin…In summary, there continues to be a lack of evidence and thus low confidence regarding the sign of trend in the magnitude and/or frequency of floods on a global scale.[25]

All the recorded evidence on these weather events shows that they are no more frequent or intense lately than they were in the past.

Like transient storm damage to the barrier reef, extreme weather events have become part of the alarmist narrative. Modern media enable everyone in the world to know about every "extreme weather event" in the world. Not long-ago tens of thousands of people in faraway countries could die in storms, floods, mudslides, earthquakes droughts and the like without the rest of the world knowing until travellers returned to tell the tale.

Another factor contributing to our perception of increasingly catastrophic weather is the fact that many people choose to live in forests and floodplains where fires and floods are well-known hazards. Wildfires in parts of the U.S. that were hardly settled in the recent past now claim hundreds of houses, including those of the rich and famous, that makes them newsworthy in their own right. In coastal areas in the less developed world the loss of life has fallen dramatically in recent times because more people live in more robust buildings.

The world is far safer today than at any time in the past and deaths from natural disasters have dropped dramatically. However, there is considerably better reporting of natural disasters and now we have cameras everywhere that can capture dramatic footage of roofs being torn off, conflagrations, floods and people in desperate circumstances. Thus, it is amazingly easy to manipulate these reports and images for propaganda and make the claim that every unusual event is more evidence of man-made global warming. How does catastrophic weather relate to CO_2? The bottom line is that a storm does not signal anything about CO_2 neither does a hurricane nor a flood. There is no plausible scientific explanation for this at all and the IPCC and many other scientists have acknowledged this repeatedly.

It is the same story with wildfires.[26] Of course there are still bushfires and wildfires, as there always have been, but they are not increasing in frequency or intensity nor has any scientific evidence been put forward to explain how anthropogenic CO_2 could possibly cause this. If mild warming of less than 1°C can cause bushfires then the Earth must have been a giant fireball when it was more than 10°C warmer than today, yet we know from the fossil records that this is not true.

What is abhorrent is how the alarmists exploit every single unusual event and use it to claim evidence of CAGW. Anytime there is a hurricane, flood, fire or drought somewhere in the World, alarmists point it out and screech "climate change!" in scant regard to the fact that these unusual events have occurred as long as the Earth has existed and there is no discernible upward trend in extreme weather.

Australia suffered deadly fires in 2009 on "Black Saturday" which killed 173 people. The cause of the fires were arson and downed power lines. The heatwave at the time contributed to the spread of the fires but did not cause them. Another major contributing factor was the heavy fuel loads. Local Councils dominated by environmentalist Greens oppose back-burning and clearance of fuel loads – the dry leaves, twigs, underbrush and other dead material recycled by nature. This of course increases the chances of wildfires since as the fire triangle shows us; oxygen and heat still require fuel. With more wildfires then The Greens can point at them and utilize the video and pictures as propaganda which "prove" global warming. This is exactly what happened in the 2019-2020 Summer when millions of acres of Australia were burnt by bushfires.

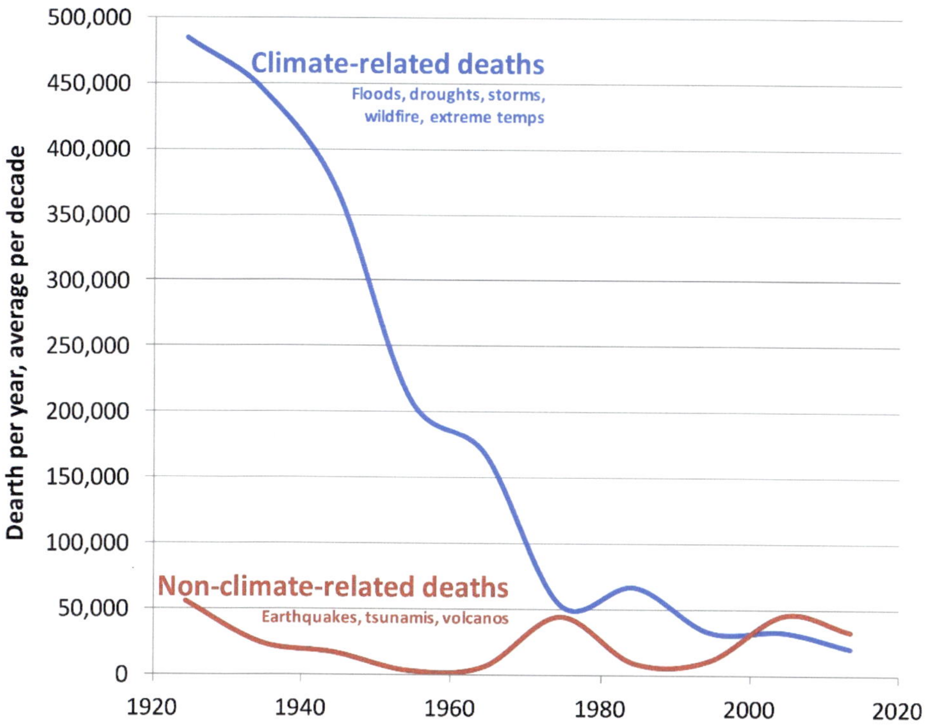

Deaths from Climate and non-Climate Catastrophes, 1920-2017
Contrary to the message of climate alarmists who claim deaths will increase from global warming, the individual risk of dying from climate-related disasters has declined by more than 90% in the past century despite the global population increasing four-fold.[27]

The Remarkably Resilient Barrier Reef

According to the official story the reef is in a very perilous state due to warming and acidification of the water caused by increasing atmospheric CO_2. The front page of The Sydney Morning Herald on Monday October 8th, 2018 proclaimed *Eliminate Coal or Lose the Great Australian Barrier Reef says IPCC.*

However, the coral of the reefs are built by countless species of organisms that evolved over millions of years to handle the fluctuating conditions. Short term local damage from storms and bleaching events is soon repaired, but the impact of those events make dramatic newsworthy pictures when they became grist for the mill of climate alarmism.

The reef grows in water with a temperature range from 25°C in the south to 27 °C in the north and what you have not been told is that any warming will most likely increase the growth rate of the coral, after all the corals in the warmer waters of the Philippines (28-32°C) grow significantly faster than the corals on our Great Barrier Reef. More CO_2 will probably increase the growth rate as well as it does for plants and indeed tests in controlled conditions revealed that many corals grow well with up to six times the amount of CO_2 in the air. Similarly prawns and shellfish can grow bigger and faster with more CO_2 in the tank.

Owing to its amazing natural beauty the reef is a convenient hostage for alarmists; "Stop carbon emissions or the reef gets it!", with it come ludicrous claims about the reef having been around for "millions of years". The GBR is in fact 8,000 years old; if you swam out to the deepest part of the reef and took a time machine back 12,000 years you would be standing on dry land. The current reef is perhaps the sixth iteration of an extensive reef system, the previous ones having each been destroyed by past dramatic falls in sea level occurring over the last few hundred thousand years.

The very slowly rising seas

Sea level rise is a tsunami of laughable prophecies, from the disappearance of islands and even countries, to the seas rising over 100 metres. Sea levels are a complex issue with multiple varying factors, not all of which are fully understood and the simplified, dumbed-down narrative of warming = melting = higher seas is peddled by the simple-minded and propagandists. We have only been able to accurately measure proper sea level changes since the 1990's when the first satellites were sent up. Once again, immediately attributing any sea level change to man-made CO_2 emissions without addressing every other possibility (e.g. land sinking or growing, thermal expansion of water, evaporation) is absurd and unscientific.

Sea levels vary regionally, for example the Bureau of Meteorology records from the gauge at Fort Denison in Sydney Harbour indicate that the sea level fell slightly in the last century from a level of 1.11 metres in 1914 to 1.05 metres in 2019. This indicates one of the most interesting features of sea levels: the situation is different in different parts of the world. In some places the land is rising and in other places it is sinking. Subsidence can cause problems but these have nothing to do with climate change and they are manageable. In The Netherlands, where a third of the nation is below sea level, the Dutch have shown just how manageable it is. The most recent scientific study of the Pacific atolls shows that 80% are either unchanged in size or getting bigger rather than smaller. Islands like Tuvalu and Kirribati that were supposed to be underwater by now are actually growing.

There is no need for alarm about flooding if the Arctic ice and other sea ice melts because it is floating (ice is less dense than water) and the water level will barely change if it melted. It was the melting of the massive ice sheets covering North America and Eurasia 14,000 years ago that caused the dramatic sea level rise of about 100 metres cutting off the land bridges between Asia, America and Australia. Humans have had to deal with changing sea level for thousands of years and we are much better at it today thanks to technology and innovation.

Gambling Fallaciously

Having been caught out so many times with failed predictions and unrealistic prophecies of doom alarmists started to broaden their predictions to encompass almost anything that they could categorise as unusual or extreme. This was facilitated by the switch from "global warming" to "climate change" so that anything that changed could be labelled as climate change and therefore attributable to man. Climate change can now cause more snow and less snow, both floods and droughts, more rain and less rain, stronger winds[28], weaker winds[29], coral reefs to die and also grow and a whole list of bizarre claims from increased acne to the taste of tea[30]. Alarmists are playing roulette by betting one chip on every number then claiming they won when their number inevitably comes up.

Chapter X
THE IMITATION GAME
The Fake Scientific Consensus

"A lie doesn't become truth, wrong doesn't become right, and evil doesn't become good, just because it's accepted by a majority."

BOOKER T. WASHINGTON

How many times have you heard the phrase "97% of scientists agree"? How about "eight out of ten owners said their cat prefers it!"? Have you ever wondered where this 97% figure came from? Not many people do and even fewer understand why. If these assertions were true it would not matter how many or how few people agreed. Truth is not determined by consensus.

There are thought to be three different sources of the 97% claim. First, the largest and most deceptive source was a 2013 report by John Cook and associates[1] claiming that 97% of climate scientists accept AGW – Anthropogenic (human-caused) Global Warming. The second was a loosely-worded web-based poll in 2009 performed by Peter Doran and Maggie Kendall Zimmerman of the University of Illinois at Chicago. The third was a short essay published by Naomi Oreskes in 2005[2].

First to be clear what AGW means. It can be used in two ways. One definition is the **dangerous AGW view** that warming is a serious problem and human emissions of greenhouse gases (chiefly CO_2) are the major cause of warming. This is known as Catastrophic Anthropogenic Global Warming or CAGW. The other is the **benign AGW view** that warming has happened (without being alarming) but the human contribution is insignificant and is nowhere near the major cause.

Many climate realists subscribe to the benign AGW view. Some realists accept that there is a contribution from CO_2 and some do not. Since realists are not alarmed about warming the question of the causes can be treated as a fascinating research question with no immediate implications for government policies.

There is another category of people called *lukewarmers* who are concerned about adverse effects of warming but not enough to support the current efforts to stop it. Author Bjorn Lomborg is the most strident advocate of this position. He considers that the benefits of warming will outweigh the adverse effects up to another two degrees of warming and all the current efforts to reduce emissions are a complete waste of money which should be spent on more useful things.

The principal source of the 97% claim was the 2013 paper *Quantifying the consensus on anthropogenic global warming in the scientific literature* (2013)[3], published in a climate science "house journal". For John Cook, the senior author of the paper, AGW means dangerous AGW and there is no hint in his paper that there is a such a thing as benign AGW. By failing to distinguish between the two, the paper was very misleading. In view of the much broader definition of AGW, that includes the benign view held by some sceptics, it is not surprising or even unusual to find a report that climate scientists accept AGW. You might as well claim that 97% of astronomers accept that the sun is the centre of the solar system. You might even find that 100% accept it without expecting anyone to be impressed.

The 97% claim is surely the most widely reported and misrepresented statistic ever. It was deemed to be the scientific paper that was most cited in the media in 2015 and this claim circulated by the prestigious international news agency *Reuters* is typical of the coverage;

> Ninety-seven percent of scientists say global warming is mainly man-made...an international study showed on Thursday. The report found an overwhelming view among scientists that human activity, led by the use of fossil fuels, was the main cause of rising temperatures in recent decades.[4]

It would have been helpful to preface the story with the warning that the nine co-authors of the study are dedicated environmental activists and climate alarmists. The study also never mentioned fossil fuels at all; this was completely made up by *Reuters*.

Cooking the Books

Many defects in the research methods have been noted in Cook's paper, the most serious being the failure to make any definition or distinction between CAGW, AGW and naturogenic global warming. This is perhaps not surprising since John Cook is not even a climate scientist. This fact is more significant when you remember that alarmists constantly denigrate any sceptic who is "not a climate scientist". What is good for the goose is good for the gander.

There were two stages in collecting the opinions of the scientists and the results were presented in a complicated way with irrelevant and distracting analysis. It is necessary to read the paper several times to be clear about what they actually found, as distinct from their personal opinions and what they want the reader to think that they found. The *Reuters* reporters almost certainly did not read the paper at all, just the media release. Cook's method was quite simple; he and his team would read the abstracts of some 12,000 published papers on climate and independently judge whether they had a position on AGW or not. Note that they seemingly read the abstracts (the brief summary at the beginning) and not the paper itself. This is entirely subjective and about as scientific as judging a movie by watching the 60 second trailer and then submitting a ranking on the *Rotten Tomatoes* website, claiming you had seen the movie.

It was not a study of scientists at large; it was a study of the climate science literature where the authors make a living from the massive funding for climate research. There are a great many sceptical scientists who do not publish in the climate science literature and they have as much relevant expertise as any official "climate scientist".

Even with all this going in their favour, the paper, as it says quite clearly in the abstract, could only find 32.6% who endorsed AGW:

> We find that 66.4% of abstracts expressed no position on AGW, 32.6% endorsed AGW, 0.7% rejected AGW and 0.3% were uncertain about the cause of global warming.[5]

So how did 32.6 become 97? Let us have a look at cat food advertisements to explain this. Just how does anyone know that eight out of ten cats prefer a particular type of cat food? Did they ask the cats? In reality, the company simply asked cat owners if their cats liked their cat food and 80% said yes. Cats like food, what a surprising discovery. Funnily enough after a complaint to the UK Advertising Standards Authority, the slogan was changed to "eight out of ten owners who expressed a preference said their cat prefers it". This does not have quite the same ring to it, but it was of little consequence. The original slogan had been seared into people's minds. Why is this particular example funny and what does it have to do with 97% of climate scientists? Well here is the key extract from Cook's paper (with our highlighting):

> Among abstracts **expressing a position on AGW**, 97.1% endorsed the consensus position that humans are causing global warming. [6]

In other words, they managed to manipulate their statistics until it gave them the answer they wanted. Even with their flawed methodology they couldn't conjure up a large enough number so they simply cooked the books. The "97%" is a statistically misleading marketing catchphrase pulled from a flawed paper. But the damage has been done. The "97% of scientists…" jingle is stuck in the heads of everyone, dutifully aided and abetted by the media who repeat it *ad nauseam*.

What does that 97% mean in relation to the questions that we want to answer to guide climate policy?

- **How much warming?**
- **Do we need to worry about it?**
- **What is the human contribution?**
- **And, most important, what is the role of CO_2?**

The answer is that it tells us nothing that we need to know. Yet President Barack Obama saw fit to use the journalistic misrepresentation of the paper to justify his support for strong action at the Paris Convention in 2015 and even added "dangerous" to the already distorted findings[7]. The word "dangerous" was never mentioned in Cook's research.

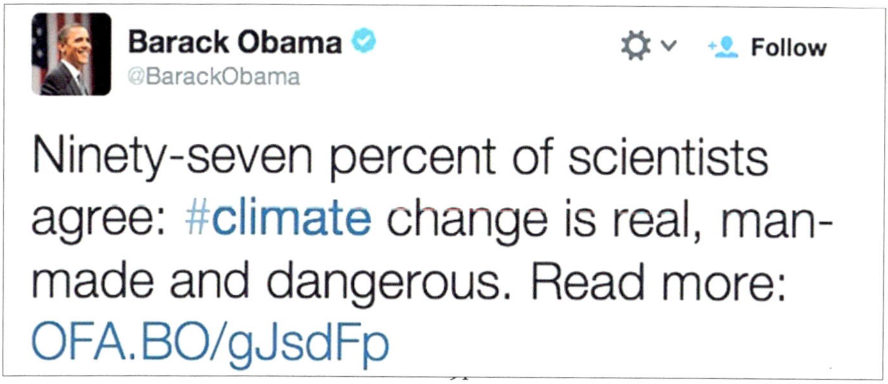

The outcome of the Paris Convention

After all the excitement of the meeting and especially the energetic contribution by President Obama, inspired by the "97% consensus", nothing much has happened. Most of the nations have done nothing, or next to nothing, or nothing that makes any significant difference to CO_2 emissions. Of course, Australia was the exception and has done a great deal to increase the cost of power and undermine the security of supply. The advice of Chief Scientists, who said that it will not make a noticeable difference to the planet, was ignored.

The 97% consensus story was played up and misrepresented to garner support for commitments in Paris but the action that has ensued around the world, with the exception of Australia, is as inconsequential as the contents of the paper itself.

Other authors have published papers with similar results to the Cook *et al.* findings which also made highly dubious claims that have since been discredited. In 2009 Peter Doran and Maggie Kendall Zimmerman sent out an online survey to scientists. Over 3000 responded and when the authors did not get the answers they wanted, they simply threw out all but 79 and claimed 77 of those (97%) were in agreement[8].

If Science is settled by consensus...

If there were a 97% consensus or even a 99% or 100% consensus it would not make any claims more valid since science is not done by consensus. If science were settled by consensus then mankind would not have made much progress. It was the shift away from "consensus" to the scientific method that fueled the Enlightenment and Industrial Revolution and the unparalleled discoveries and innovations that have shaped the modern world. When we look back at many claims that were the overwhelming consensus: geocentrism; miasmas being the cause of disease; fixed continents; ulcers caused by stress, just to name a few, we often find it was one or two scientists versus the "consensus". Often this consensus vehemently and sometimes violently opposed anybody proposing new or alternative theories who were often branded sceptics and deniers.

Perhaps the most infamous example is that of Galileo Galilei who was convicted of heresy for postulating the theory of heliocentrism. "In questions of science, the authority of a thousand is not worth the humble reasoning of a single individual", said Galileo and eventually his humble reasoning defeated the consensus.

Alfred Wegener's theory of continental drift was mocked for decades, critics routinely pointed out that Wegener was not a geologist. When Einstein published the Theory of Relativity German newspapers reported "One hundred German physicists claim Einstein's theory of relativity is wrong." Einstein, echoing Galileo's words, purportedly remarked, "If I were wrong, it would only take one."

Even Einstein was sometimes wrong, most notably about the cosmological constant. Yet, unlike most alarmist scientists, Einstein admitted his error when Georges Lemaître and later Edwin Hubble proved their theory through observation and experimentation. A scientist is not true to science unless they can accept evidence that proves their theory to be incorrect and then adjust their theory or throw it out altogether. Though new and alternative theories are not always in themselves exact, they often lead scientists on a journey to better theories; Wegener's continental drift theory eventually led to the discovery of plate tectonics.

[9]

Science is not a vote; the Scientific Method was first formalized in the 17th Century precisely to ensure this. It formulated a series of steps that started with a question or observation, forming a hypothesis, experimentation, recording and checking results and finally drawing a conclusion and reporting the results.

[10]

Consensus, on the other hand, is important in politics and society. It plays a pivotal role in Asian cultures but it has no place in science. Best-selling author and filmmaker Michael Crichton (*Jurassic Park*, *ER*, *Westworld*) made this excellent point in a 2003 speech called "Aliens Cause Global Warming":

> I want to pause here and talk about this notion of consensus, and the rise of what has been called consensus science. I regard consensus science as an extremely pernicious development that ought to be stopped cold in its tracks. Historically, the claim of consensus has been the first refuge of scoundrels; it is a way to avoid debate by claiming that the matter is already settled. Whenever you hear the consensus of scientists agrees on something or other, reach for your wallet, because you're being had.
>
> Let's be clear: the work of science has nothing whatever to do with consensus. Consensus is the business of politics. Science, on the contrary, requires only one investigator who happens to be right, which means that he or she has results that are verifiable by reference to the real world. In science consensus is irrelevant. What is relevant is reproducible results. The greatest scientists in history are great precisely because they broke with the consensus. There is no such thing as consensus science. If it's consensus, it isn't science. If it's science, it isn't consensus. Period.[11]

A petition signed by over 31,000 American scientists circulated in 1998 and 2007 stated this:

> There is no convincing scientific evidence that human release of carbon dioxide, methane, or other greenhouse gases is causing or will, in the foreseeable future, cause catastrophic heating of the Earth's atmosphere and disruption of the Earth's climate. Moreover, there is substantial scientific evidence that increases in atmospheric carbon dioxide produce many beneficial effects upon the natural plant and animal environments of the Earth.[12]

Most people have never heard of this petition. Climate realists do not need to mention or publicise it as they are aware science is not settled by consensus. It is usually only ever mentioned in response to the "97%" claim but rarely do realists lead with it, they prefer the science to speak for itself.

The Consensus Refugees

The critical need for a consensus on the global warming theory has produced a dark underside. Many of those who disagree are subjected to bullying or find themselves ostracised. Others find their careers and livelihoods destroyed. Many who dissent, are sceptical or are whistle blowers find themselves cast out of their professions and unfairly accused of dishonesty and treachery. Professor Peter Ridd, for example, spent much of his career at James Cook University (JCU) in Queensland researching the Great Barrier Reef. He achieved an international reputation but was dismissed in 2018 for pointing out that some of his colleagues were misrepresenting the state of the reef and their results needed to be checked more thoroughly and replicated. Ridd was proven correct when a paper revealed that eight studies from JCU could not be replicated[13] and one researcher was found to be fabricating data[14]. That researcher had no action taken against her by the University, but Ridd, the whistle-blower, was fired for raising concerns.

The same University also treated the late Professor Bob Carter, who had worked there for 31 years, with disdain for his outspoken scepticism. As an author, speaker and advisor to Governments Professor Carter was, like Peter Ridd, upsetting the money-making machine at JCU. So, the University took his office, his title and then, ramping up the spite, the University even took his library card and email account.

There are many other refugees from the proclaimed consensus who have been isolated, sneered at, mocked, branded as illegitimate and deemed as unworthy of remaining in the hallowed halls of academia. Others keep quiet until they have retired from academia and then speak up, coming out as sceptics. The atmosphere of fear is rife and it is understandable that some scientists do not want to lose their jobs as they have families to feed. Those that do speak out are either tenured or very brave.

How can the apparent consensus scientists claim "the debate is over" when they have silenced or fired climate realists? In fact, no debate has ever really taken place because they did not wish one to be held. Instead, they resorted to devious tactics and behaviour that shames their profession. Once you are branded a heretic there is no need for your critics to respond to your arguments or justify their vicious attacks against you.

The very shrillness of the response to any scepticism is itself a telling point. If the case for man-made global warming is so strong then why would there be a need to manipulate, distort and invent a consensus, especially when science has nothing to do with consensus? Let the theory, evidence and observation speak for themselves, why lie? Furthermore, why seek to silence those that hold alternate theories and ideas or point out flaws, errors and weaknesses in the case for CAGW?

The fake 97% consensus is a means of softly cajoling people into assenting to a lie and silencing those who try to call it out. Some alarmists have called for "climate change denial" - even though there is no such thing - to be a crime and "deniers" to be imprisoned and even executed[15]. History is replete with examples of people being silenced, censored, exiled and executed for their thoughts and ideas. It has been most prevalent in societies that are totalitarian or religiously fundamental in nature. The 97% consensus myth is not just some misleading statistic, it miseducates the public about science itself by implying that science is a vote and that a majority can silence and punish a minority. This is not about science; it is about politics and power.

Chapter XI
APOCALYPSE NOW
The Propaganda of Alarmism

"The point of modern propaganda isn't only to misinform or push an agenda. It is to exhaust your critical thinking, to annihilate truth."

GARRY KASPAROV

Propaganda is a tool used to persuade and motivate. That does not necessarily mean it is always bad or deceptive. Advertisements for weight loss or quitting smoking or gambling are propaganda and so are ones pushing you to consume beer or McDonalds. Propaganda is not very concerned with truth or facts, it is about eliciting an emotional reaction and persuading the target to believe something. A good way to understand if you are being propagandised; if your initial reaction is emotional, it's probably propaganda. Sadly, truth and facts, especially in the realms of science, are often boring to most people which is why propaganda in science is so dangerous; the bland independence of unadulterated science cannot compete with the entertainingly deceptive propaganda of climate alarmism.

Not enough people have the training and the time to form an independent opinion on climate issues and so most follow what appears to be the majority opinion, or expert opinion, or simply the narrative that captures the public imagination. The young are especially vulnerable to propaganda as their reasoning skills are not fully developed and they are usually fed a very one-sided view at school. This chapter challenges the received narrative of climate alarmists.

Most people are low information consumers of climate science news. They are often unable to answer such basic question such as the actual amount of CO_2 in the atmosphere or the mechanism of the "greenhouse effect." Yet, they can certainly rattle off lists of logical fallacies such as "climate change means bad weather and therefore bad weather means climate change is real" or "97% of scientists agree" or "CO_2 is a pollutant". And they can certainly make their "arguments" very passionately, yet such people are merely following Galileo's maxim on matters that require critical thinking. [1]

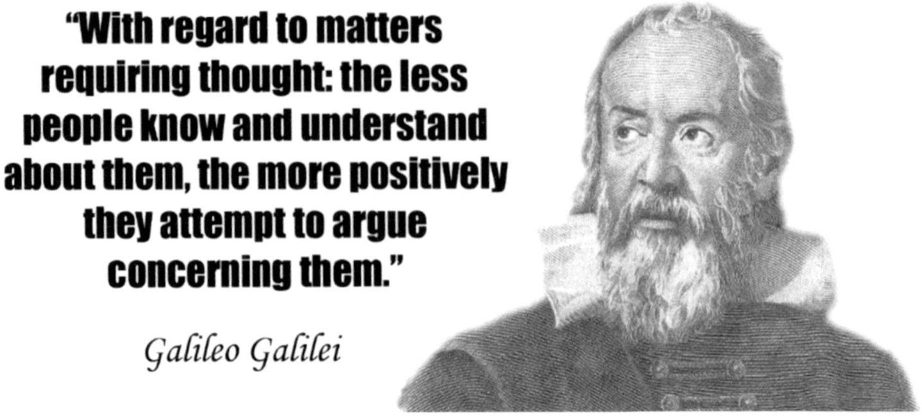

The Art of Propaganda

Every propaganda technique is used to promote the global warming agenda in a manner that would make a totalitarian state proud. The early 20th Century saw the development of imagery as propaganda, first in photographs and then moving pictures and this has been utilised very effectively by the global warming propagandists. The technique of climate alarmists is to show emotive images of bleached coral or stranded polar bears, often with ominous background music.

Images can have massive and lasting subconscious impact even after we know in the thinking part of our brain that they are not telling the truth. Climate stories in the press are often illustrated with pictures of smoking chimneys. Clearly the purpose is to convey the impression that carbon dioxide is dirty and therefore bad[2].

What is going on here? Well firstly the "black smoke" is steam, which is water vapour if it is coming from a power station. It is not "carbon pollution" and it is not CO_2 because if it were you would not be able to see it. Even primary school children should know that carbon dioxide is colourless. If the black smoke is supposed to represent carbon dioxide, then the picture is telling a lie. The steam from cooling towers is white, however by taking the picture when it is backlit (against a sunset) or digitally altering the picture the white steam appears to be black.

The Pocerady coal fired power station in the Czech Republic taken during the day and backlit at sunset. The "smoke" is steam, like that rising from a boiling kettle.[3]

Other imagery used is usually entirely devoid of context or selectively used to appeal to emotion. Bleached or damaged corals are always blamed on climate change when it could have been any number of factors – a cyclone, the Crown of Thorns starfish or just the natural die-off and regeneration of the coral. Sad looking animals or creatures in distress is another favourite subject and all manner of spurious leaps are taken to describe how it is climate change that is the reason a koala has to travel further to find eucalyptus leaves or the hippopotami of the Serengeti are distressed because of a changed habitat. Animals can now, apparently, express emotions readable by humans.

The Language of Propaganda

A central theme of George Orwell's *1984* was the changing of language which, when used in a maliciously political way, could manipulate and control people. He would not be surprised to see what is occurring today. The language of alarmism is being ramped up, as though it was not sufficiently emotive already. "Global Warming" was rebadged as "Climate Change" after satellite measurements over the past two decades showed that the Earth has not actually warmed. True to the *1984* script the historical record had been revised. Any weather event could then be cast as proof of climate change – a hot summer, a cold winter, a drought, a flood, a storm, a bushfire and it is attributed to man-made CO_2. A global warming trend can be refuted by reliable records but natural climate change cannot be refuted because it has always been going on and nobody denies that.

Yet even that was not enough. In the 'doubleplusungood' manner of *1984* "Newspeak" *The Guardian* newspaper issued new guidelines for reporting on the environment. "Climate change" was no longer emotive enough and preferred terms are now "climate emergency, crisis or breakdown" and "global heating".

> We want to ensure that we are being scientifically precise, while also communicating clearly with readers on this very important issue," said the editor-in-chief, Katharine Viner. "The phrase 'climate change', for example, sounds rather passive and gentle when what scientists are talking about is a catastrophe for humanity.[4]

Other terms now used are "global heating" and "climate collapse". Greta Thunberg, the Swedish teenager who has encouraged children to strike for climate, laid out a smorgasbord of fear-laden phrases in a viral Tweet[5].

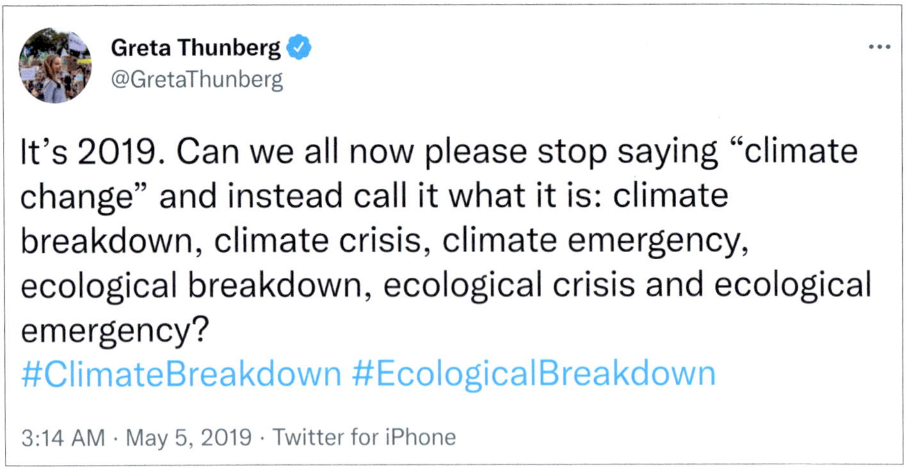

Greta Thunberg
@GretaThunberg

It's 2019. Can we all now please stop saying "climate change" and instead call it what it is: climate breakdown, climate crisis, climate emergency, ecological breakdown, ecological crisis and ecological emergency?
#ClimateBreakdown #EcologicalBreakdown

3:14 AM · May 5, 2019 · Twitter for iPhone

Demonizing the "Other"

Whilst emotional language can be used to amplify global warming fear the next key technique of propaganda is name-calling and stereotyping of the "other" – in this case the persistent denigration of "science deniers", "climate deniers" or "climate change deniers". The use of this language in the climate science debate is dishonest and casts grave doubt upon the credibility of all the people who use it. This is the verbal pollution used to bring contempt upon people and subtly likening them to those who deny the Nazi Holocaust.

Many internationally recognized climate scientists are not alarmed about warming and do not think CO_2 is a dominant influence in the warming that has occurred. They make their living out of studying climate and know that it changes because that is the central issue in the field. All those who take the trouble to publicly challenge the narrative of alarmism are interested in the scientific issues, many are professionally qualified and know enough about the issues to form independent opinions even if they do not publish in journals.

Look at the highly qualified scientists who are sceptical of some or all the alarming claims, including Richard Lindzen, Will Happer, Judith Curry and lately Steven Koonin. And in Australia the list includes the late Bob Carter, Bill Kininmouth, Ian Plimer and Garth Paltridge. Using the language of Holocaust denial to denigrate these scientists destroys the credibility of the stone-throwers. That is the language of the primary school playground, not an adult discussion.

It makes no sense to talk about climate deniers, climate change deniers and climate science deniers. You cannot deny science because science is an on-going process not an end-state. Those who fling this accusation around clearly do not understand how science works. Perhaps you can deny the scientific *method* which, ironically, is what many of the global warming alarmists very often do! Similarly, nobody denies climate change, everyone knows that the climate changes. The record of climate change is clear as far back as we can tell. There were major Ice Ages and only a few centuries ago there was the Little Ice Age. After the Ice Ages the Earth warmed and life flourished under much warmer temperatures than we have today and with far more carbon dioxide in the air.

In fact, the term "climate change" is really a tautology, the "change" is superfluous, climate always changes, that is what it does otherwise it would not be climate. It is like saying "evening sunset" or "sad misfortune". The term "climate change" gives the impression that climate should be fixed and unchanging and only man changes it by using coal and oil.

The demonization of the "other" does not just apply to people or ideas. Terms such as "carbon pollution" and "dirty coal" are now ubiquitous. What is "carbon pollution", apart from the dishonest image of black smoke pouring out of chimneys? Children who know that CO_2 is colourless will wonder what the adults are up to when they talk about "carbon pollution".

Further, school children with a grounding in the basics of chemistry should know that 'carbon' is not 'carbon dioxide'. **Carbon** is an element in the periodic table and a solid at room temperature. **Carbon dioxide** is a compound and a gas at room temperature. Carbon is the second most common element (18%) in the human body after oxygen (65%). Life on Earth would not exist without carbon, it is the building block of every living organism on the planet. Are school children even taught these basic facts? Do they know what their bodies are made of? Or that CO_2 is plant food? To describe the basis of organic matter and life on earth as "unclean" is a sign of scientific illiteracy.

Carbon itself is not "dirty". Here is a picture of pure carbon, beautiful isn't it? Of course, if you still think carbon is dirty and you have any diamonds then you should send them to the author immediately.

Diamond is a pure form of carbon. The element occurs in millions of compounds, is the sixth most common element in the universe, the fourth most common element in the solar system and the second most common element in the human body.[6]

Of course, carbon dioxide is not the only thing emitted from smokestacks and car exhausts, some of the other compounds, such as sulphur dioxide, are genuine pollution and they have been largely controlled by technological innovation and strict regulations.

The Ministries of Propaganda

Propaganda always requires a vector and climate alarmism has three main carriers; the media, education and the entertainment industry.

The mainstream media are the bull to the red rag of propaganda, enthusiastically and hyperactively charging headlong into every story. The main issues with the media are: bad news sells, politicisation, lack of scientific literacy and lack of professionalism and failure to observe codes of ethics in reporting. It is one of the cliches of reporting that bad news attracts sales and clicks. As they say in the newsroom "If it bleeds it leads" and climate news has been a leader for a long time. Often, they will publish propaganda that not only contradicts previous pronouncements but all logic and common sense. Searching back through news archives it is easy to find published predictions that never materialised; warnings that the planet only has two decades, fifteen years, ten months or five weeks to save itself. These deadlines come and go without so much as an admission of error before the next Doomsday prediction is rolled out.

Over the past decade reports have cropped up in nearly every major country that their locale was *Warming Twice as Fast* (WTF!)[7] as everywhere else. How can everywhere be warming twice as fast as everywhere else?! This is next level propaganda on a global scale. The old adage that: *A lie gets halfway around the world before the truth has a chance to get its boots on* has never more been aptly demonstrated.

Warming Twice as Fast (WTF)...everywhere!

SCIENCE

Arctic Is Warming Twice As Fast As World Average

December 18, 2014 · 3:36 AM ET
Heard on Morning Edition

CHRISTOPHER JOYCE

Technology & Science

Canada warming at twice the global rate, leaked report finds

Government report scheduled for release on Tuesday

CBC News · Posted: Apr 01, 2019 3:17 PM ET | Last Updated: April 5, 2019

FEATURED, LIFE, NATIONAL, STATE NEWS, WORLD

New Jersey warming twice as fast as globe

by Staff Report • September 24, 2015 • 0 Comments

Parts of United States are heating faster than globe as a whole

A new study shows the Northeast USA will reach the dangerous 2°C warming threshold faster than most of the rest of the planet

Adirondacks warming faster than the global average, study shows

Gwendolyn Craig Oct 27, 2018 2

IPCC: Europe has been warming faster than the global average

UN climate science panel's full report shows how climate change could pan out differently across the many regions of the world

Britain warming faster than average

Britain's climate is warming faster than the global average quoted in the latest IPCC report, according to climate experts.

CLIMATE CHANGE

Himalayan meltdown: Hindu Kush Himalayan region warming faster than global average

Capping global warming at 1.5°C will still mean a higher rate for the region through this century

NEXT NEWS ›

By Richard Mahapatra
Last Updated: Monday 04 February 2019

Tibet is warming at twice global average

EARTH 24 July 2007

By Phil Mckenna

Science News *from research organizations*

American West Heating Nearly Twice As Fast As Rest Of World, New Analysis Shows

Date: March 30, 2008

Source: Natural Resources Defense Council

MOUNTAINS ARE WARMING TWICE AS FAST AS THE REST OF THE WORLD

To sustain mountain communities, we need more monitoring and better governance.

BOB BERWYN · JUN 19, 2017

Northern Hemisphere Warming Twice as Fast as South

15 DECEMBER 2006

Singapore warming faster than average

ENVIRONMENT JANUARY 28, 2015 / 8:25 AM / UPDATED 6 YEARS AGO

Australia temperatures rising faster than rest of the world: official report

By Jane Wardell 3 MIN READ

News Release 12-232

Study Finds That Portions of the West Antarctic Ice Sheet Are Warming Twice as Fast as Previously Thought

Findings could have important implications for global sea-level rise

Europe's North Sea Warming Twice as Fast as World's Oceans

By lenrosen4 September 9, 2017 8

The Surface Of Oceans In The Southern Hemisphere Is Warming Twice As Fast As We Thought

CHRIS PASH
OCT 6, 2014, 4:00 AM

Korean Peninsula is warming at a speed faster than global average

Posted on : May.8.2009 10:47 KST Modified on : May.8.2009 10:47 KST

Finland is warming faster than the rest of the world

China's heating up twice as fast as the rest of the world

Sweden's temperature is rising more than TWICE as fast as the global average

DECEMBER 25, 2015

Russia warming '2.5 times quicker' than global average: ministry

by Maria Antonova

APRIL 13, 2010

Spain warming faster than rest of northern hemisphere: study

Norway warming fast

PUBLISHED APRIL 15, 2019 · UPDATED APRIL 15, 2019

Norway's rate of temperature increase is nearly double the global average

Report: Alaska warming twice as fast as global average

By Travis Khachatoorian
Published: Aug. 9, 2017 at 3:14 PM GMT+10

National Parks are warming twice as fast as the U.S. overall

Climate change poses the greatest threat to some of the country's most prized areas.

Both the media and the climate debate have split along party lines. Nowadays most of the mainstream media outlets are strongholds of left-wing orthodoxy and also climate alarmism while platforms that allow space for the other side of the case are shrinking. Hardly any journalists have any training in science and they have to write with little or no understanding of scientific matters. Those who are trained mostly follow the "science" of alarmism, or pretend that they do, or just report climate stories without comment. The resulting coverage is repetitive and one-sided, and it does nothing to lift the level of science literacy in the community or among politicians.

Basic principles of good journalism, like checking the facts, are habitually abandoned in reporting climate-related stories. Some of this can be attributed to ignorance of science, some to the pressure to get ahead of the opposition with breaking "news", some to incompetence and ultimately it reveals a lack of concern about the ethics of the trade at the top of the organization.

Education? No, thought control.

The education system has been shamelessly hijacked by the alarmist agenda to capture and convert young, impressionable minds. Children are being fed a diet of fear from ages that are far too young to process or question complex science by teachers who understand little of it themselves and often also have a political agenda. This is where the propaganda technique of glittering generalities is employed which can be seen in the vague, sweeping statements about saving the world and stopping environmental destruction on the placards carried at rallies and protests. In this manner children have become the consumers, disseminators and willing hostages of the alarmist propaganda machine.

If just one person from a private company presented a talk to schoolchildren extolling the benefits of nuclear power or gas there would be uproar, yet the alarmist propaganda is allowed free reign to run rampant in schools and Universities. Worse still, climate alarmism is all paid for using taxpayer's money. People are literally paying to have their children terrified and indoctrinated.

A vast web of Government-funded organisations, NGO's and groups are involved. *Sustainable Schools NSW* proudly talks about collaborative efforts "between young people and representatives from over 70 environmental and youth organisations, schools, universities and local governments"[8].

It works with the Government-funded *Office of Environment and Heritage* and is managed by the *Australian Association for Environmental Education* which lists the NSW Government *Environmental Trust* as one of its partners. They provide tips and tricks for securing money from the *Eco Schools* grants and local councils, all funded by taxpayers. These webs of activists are a deep swamp feeding on an even deeper trough of taxpayer's money.

New Jersey provides a glimpse of a possible future when it became the first U.S. State to make Climate Change education mandatory in 2020 at every grade level and across multiple subjects including visual and performing arts and physical education[9]. Children cannot even throw a ball around without being subject to alarmist propaganda. Meanwhile the performance of school students appears to be falling in Australia and many Western countries, especially in science and mathematics.

People who hate coal and oil need to appreciate how many of the things we have and hold come from the petrochemical industry. Start with your mobile phone and your sneakers, then go on to hot water, cosmetics and pharmaceuticals. The students who follow Greta Thunberg and protest in the streets should show their commitment to the cause by throwing away their smartphones, Xboxes, PlayStations and stop riding in the SUV and jet airliners.

Pulp and Poliwood

Literature has had a longer and more pervasive influence and paved the way for the environmental cataclysm narrative. The 1958 misanthropic novel *The Dance with The Devil* by former Nazi, eugenicist and extreme environmentalist Günther Schwab, sold over a million copies worldwide. The novel contained the first fictional catastrophic prediction of man-made CO_2 emissions causing seas to rise and ice caps to melt. Following closely was Rachel Carson's influential anti-pesticide book *Silent Spring* in 1962 and then in 1968 *The Population Bomb* by Paul Ehrlich and John Holdren. These books shared the common theme that man could only exist in harmony with nature by following a "look but don't touch" approach lest upsetting some ill-defined mystical balance would tip the Earth into an apocalypse. Apparently, humans are the only species that should not interact with their environment. The thread running through all these early works was overpopulation and the nascent greenhouse effect wove itself into the fictional tapestry. The 1966 sci-fi novel *Make Room! Make Room!* about overpopulation was the basis for the 1973 movie *Soylent Green*, the first movie to present global warming in a dystopian setting along with pollution, poverty and euthanasia[10].

Certain totalitarian regimes were early adopters of celluloid as a means of propaganda but global warming has taken things further by utilising not only movies but the entire entertainment institution. And what institution is better suited to propaganda than politicised Hollywood? 'Poliwood' has incredible financial and marketing resources and a cast of thousands of actors with incredible skills at turning fantasy into reality. These movie and media stars are looked up to and admired by millions around the globe, in particular the young, who hang off their every word and will buy a product merely because it features a famous face.

One of the greatest influences Hollywood has on the climate debate is the use of celebrities-as-testimonial, a propaganda technique also ubiquitous in advertising. Actors such as Leonardo Di Caprio, Russell Crowe and Kate Blanchett regularly lecture the public from their yachts and private jets on the evils of fossil fuels and imminent danger of catastrophic climate change even though they know little to no science. As comedian Ricky Gervais wittily observed, most Hollywood stars "have spent less time in school than Greta Thunberg"[11].

Pavlov's People

We are all susceptible to propaganda because intelligence and higher education do not help when our critical faculties are overwhelmed by emotionally charged pictures and grand narratives supported by a mass of public opinion and peer pressure. The combined effect of all the influences sketched above is a wall of uninformed public opinion that is hard to penetrate with tools as feeble as evidence and logic.

The repetition of trite slogans (*extreme weather, carbon pollution, dirty coal, tipping point, climate action, climate justice*) has a Pavlovian effect on people. Almost everyone is susceptible to propaganda, no matter how intelligent they may be. Conflating every negative event with climate change has hypnotised docile believers into barking and drooling every time the alarmist bell is rung. Every forest fire, cyclone, hot day, melting snowflake and lightning strike is all because of man-made CO_2 and that is "settled science" and that is what you will believe.

The most dangerous development is the cancellation of dissenting voices – in the name of "science!" No true scientist would seek to omit facts, demonise colleagues or appeal to an apparent authority or consensus. That is the way of cults, not the way we learned to think about science in modern times.

Here is a question for alarmists. If the case for concern has a genuine scientific basis, why the need for propaganda, cancelling, indoctrination and all the other deceptive practices used to propagate the CAGW doctrine? A scientific theory should be strong enough to stand on it's own two feet, support itself and defeat any opposing theories in a fair debate. It certainly should not be kneecapping the competition then declaring itself the winner.

Chapter XII
THE STING
Follow the Money

"When plunder becomes a way of life for a group of men in a society, over the course of time they create for themselves a legal system that authorizes it and a moral code that glorifies it."

FREDERIC BASTIAT

In 2015 global spending on climate change amounted to U.S. $1.5 trillion according to research conducted by the *Climate Change Business Journal*[1]. That is $1,500,000,000,000. Hard to fathom isn't it? So, let us visualise it. Imagine a pallet stacked with $100 bills as tall as a room. Now imagine them covering an area about the size of one and a half football pitches. That is how much is spent on climate change each year across these segments; renewable energy; carbon capture and storage; energy storage; energy efficiency; green buildings; transportation; carbon trading; climate change adaptation; and consulting and research[2].

In the infographic[3] the red area is the 2018 annual profits of Exxon, the purported terrible polluting behemoth. Grey represents the annual profits for Apple, the most profitable company in the world. Exxon's profit is slightly less than the cost of the Manhattan Project, adjusted for inflation. This was the three-year program that split the atom and built the first atomic bombs. The Manhattan Project, which also paved the way for nuclear power, was the largest and most complex scientific programme until the 1960's.

In 1962 President John F Kennedy announced the plan for the U.S. to land man on the moon within the decade. The twelve-year Apollo program resulted in eleven manned space flights, including seven Moon missions, employed nearly half a million people and created innumerable scientific breakthroughs and achievements. The cost of the entire Apollo program amounted to some $150 billion in today's money[4]. The amount spent on Global Warming each year is ten times the entire twelve-year Apollo program! It is by far the largest, most costly venture in the world by a gargantuan margin and in the history of mankind. The Apollo program was one giant leap for mankind at a fraction of the cost.

Imagine if we were sending over 70 missions to the Moon every year, landing there every five days, there would surely be outrage and calls to stop wasting money. The $1.5 trillion is also thought to be an underestimate as it did not include several other areas and outgoing payments such as salaries. This and the increased spending over the past five years would surely push this figure closer to $2 trillion dollars. To put that into perspective, the combined GDP of 53 of the 54 countries in Africa was just under $2 trillion dollars in 2019. If this money were shared amongst the citizens of the 30 poorest sub-Saharan African countries such as Rwanda, Sudan, Namibia and The Congo, their income would quadruple every year.

Annual Spending on Climate Change

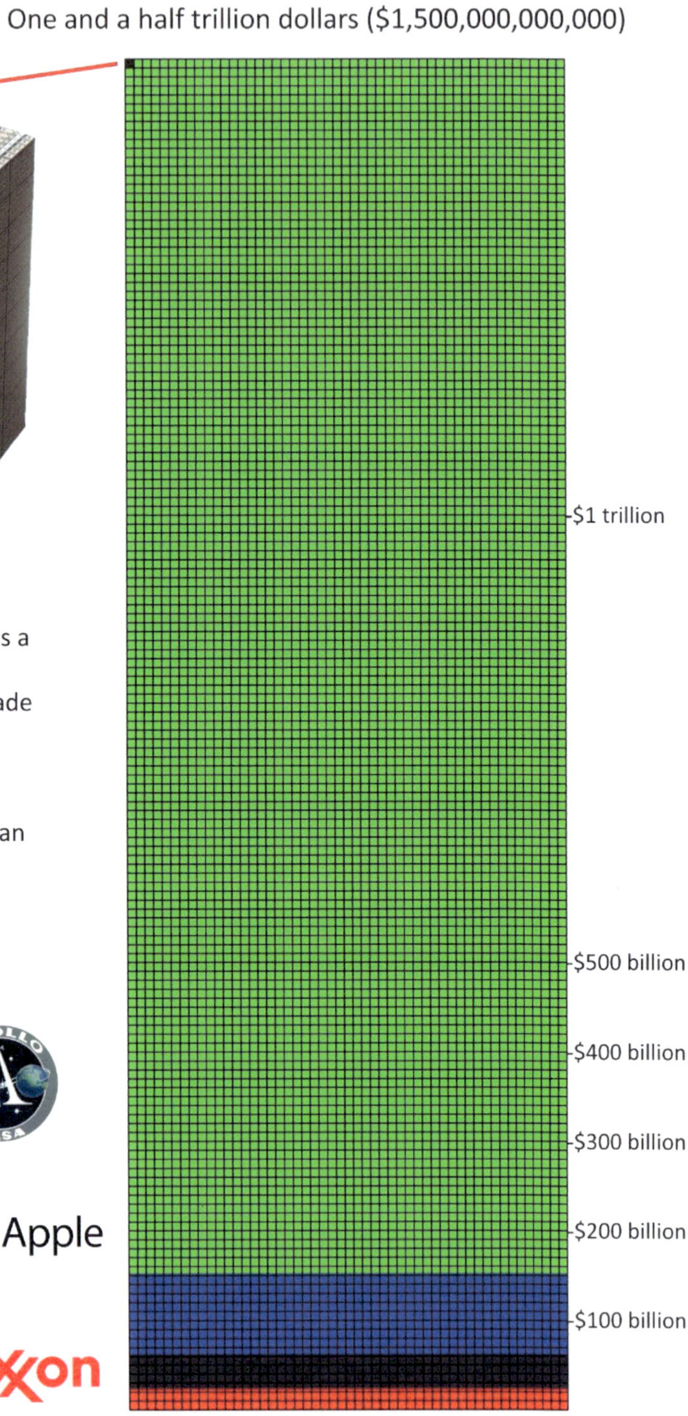

Each block is the size of a standard pallet and as high as a typical room. The block is worth $200 million and is made up of U.S. $100 bills.

The $1.5 trillion spent on climate change would cover an area greater than one and a half soccer fields.

$150 billion: The cost of the entire Apollo Program

$60 billion: The annual profit of Apple in 2018

$20 billion: The annual profit of Exxon in 2018

And this unbelievable level of spending has achieved virtually nothing. The alarmists tell us that despite this massive expenditure the Earth is getting warmer and we are still facing catastrophe. Power prices are the highest they have ever been in many countries; renewables provide a tiny percentage of unreliable power and blackouts are increasing. Global warming spending appears to be dragging us backwards to the Stone Age and making millions poorer.

The Green-Eyed Monster

Climate alarmists say that those who oppose their position receive large amounts of money from business interests with a self-interested agenda. The implication is that climate sceptics have a conflict of interest, and the views and data presented by these opponents of climate alarmism are tainted and are therefore not worthy of being refuted. The further implication is that climate alarmists, on the other hand, are free of any taint or bias and that they are only motivated by selfless concern for humanity and the environment. Naomi Oreskes is the chief academic promoter of this argument and in her 2010 book *Merchants of Doubt* she depicts struggling climate scientists, standing like David against the Goliath of science deniers, lavishly funded by Big Oil and other corporate interests. Oreskes reported that Exxon Mobil contributed $75,000 to the free enterprise Cato Institute and sums of $40,000 and $50,000 to other organizations for research challenging the consensus. She found other grants from Exxon to sceptics amounting to some $16 million between 1998 and 2004. This is roughly a briefcase full of money over a six-year period. Just compare that to the $1.5 trillion in the Annual Spending on Climate Change diagram.

Then it seems that the Big Oil money for sceptics dried up and ExxonMobil granted $100 million to Stanford[5] for research to develop renewable energy sources. It is hardly surprising that Big Oil and most of Big Business has "gone green" to chase the trillions of dollars that are on offer in the carbon credit system. Subsidies are provided and so is preferential access to the grid. Oreskes continued her pursuit of Exxon Mobil and, together with Harvard academic Geoffrey Supran, was later found to have produced flawed studies that were inaccurate, unreliable and biased.[6]

It is significant that critics like Oreskes and all those who accuse deniers of being funded by supposedly nefarious interests, automatically assume that those scientists would produce false or misleading research. It is common for private companies to sponsor research into their own products; who else will? There is no requirement or incentive for scientists to hide or dilute negative findings. In fact, it is precisely why these companies hire the scientists in the first place – they want to know if their products are dangerous or have flaws that need to be fixed.

Any researcher who falsified research or hid information would not be hired again and could severely damage their reputation. Yet the alarmists have spun a false notion that funding renders a researcher incapable of analysing facts fairly and honestly. Since their findings are in the public domain, then any scientist can review and easily refute or dispute this research. That is the process of science. However, Oreskes *et al* do not challenge the science and research produced by "deniers", they challenge who is funding it. This kind of *ad hominem* is attacking the messenger, not the message. This is not how science is done; it is pure propaganda and wholly unscientific.

However, these tenets do not apply to science in the global warming sector. No matter how often and how comprehensively their predictions fail and their research is debunked, the alarmist scientists continue to receive unprecedented levels of funding and given greater exposure and airtime. In fact, the more catastrophic and shocking (and thus more erroneous) they are, the wealthier they become. Reputations are not at risk and their rewards for promoting the alarmist line are substantial. Evidently, there is no mechanism, checks nor balances to ensure poor work; bad science and failure are discouraged within the alarmist scientific community.

Cheques and Imbalances

Whilst alarmists shriek "funded by Big Oil!" to anyone who received some loose change from Exxon two decades ago, the reality is that it is actually "Big Green" that is spending lavishly on funding alarmism. There are detailed accounts of huge donations to Green organizations from major charities, foundations and individuals. Writer Matthew Nisbet exposed the huge amounts of money going to Green organisations in a major climate science journal *Wires Climate Change*[7]. While alarmists deplore the $7 million annual budget of the free-market think tank *Heartland Institute*, it turns out that the environmentalist *Sierra Club* alone received almost $US50 million between 2011 and 2015 and the top 20 recipients of climate-related charity took a total of $230 million during that time.

Along with the Sierra Club came a plethora of supposedly environmental organisations such as the *Alliance for Climate* and *The Wildlife Conservation Society* and other seemingly innocuous sounding organisations such as the *Sustainable Markets Foundation* which is funded by the likes of The Rockefeller Family and the billionaire environmentalist Tom Steyer. These foundations are established as tax exempt charities and then simply shuffle money to other organisations. *Sustainable Markets Foundation* has lobbied against fracking, diverted money to radical causes and pushed for Government regulations to restrict exploration for natural resources[8].

Granting and re-granting bodies such as the *Energy Foundation*, are at the core of a complicated web of organisations. They appear independent but mutually support each other in many ways. The funds they receive come from a small cabal of donor organisations that crop up repeatedly; The *Rockefeller Foundation*, The *Pew Charitable Trust*, The *Hewlett Foundation* and The *Packard Foundation* to name a few. All these Foundations have been funding dubious activities such as *Planned Parenthood* for decades. David Packard initially intended his funds to be primarily concern about population control, believing population increase would mean utter chaos for humanity. "We must support abortion and any other policy that will help," Packard wrote[9].

Certain individuals have profited very nicely from Global Warming advocacy. Exhibit A is undoubtedly Al Gore who was worth a paltry $2 million when he left the office of Vice-President in 2000 and is now worth an estimated $300 million[10]. This did not come from DVD sales of *An Inconvenient Truth*.

Gore invested in a car company which had over half a billion of taxpayers money poured into it and a variety of other businesses which he knew he could profit from because he had insider knowledge. Similarly, senior Democrat Nancy Pelosi lives in a palatial mansion probably thanks to deals like the one from which her husband reportedly profited in 2014 when he purchased shares in a Green energy company shortly before an acquisition which sent its stock price soaring[11].

The nexus between Big Government and Big Warming is now so strong that two major players in the global warming industrial complex put their hands up to run for President of the United States in 2020. In 2017 billionaire Michael R. Bloomberg, the U.N. Secretary-General's *Special Envoy For Climate Change* and former Mayor of New York, announced a $50 million commitment to move nations away from coal dependence. This came after a commitment of $64 million to the *Beyond Coal* campaign in the U.S. where his efforts resulted in more than 50% of the United States' coal plants being shut down since 2011[12]. Bloomberg is unlikely to have much effect on the coal industry worldwide because many nations in the developing world are busy building hundreds of new coal-fired power stations[13].

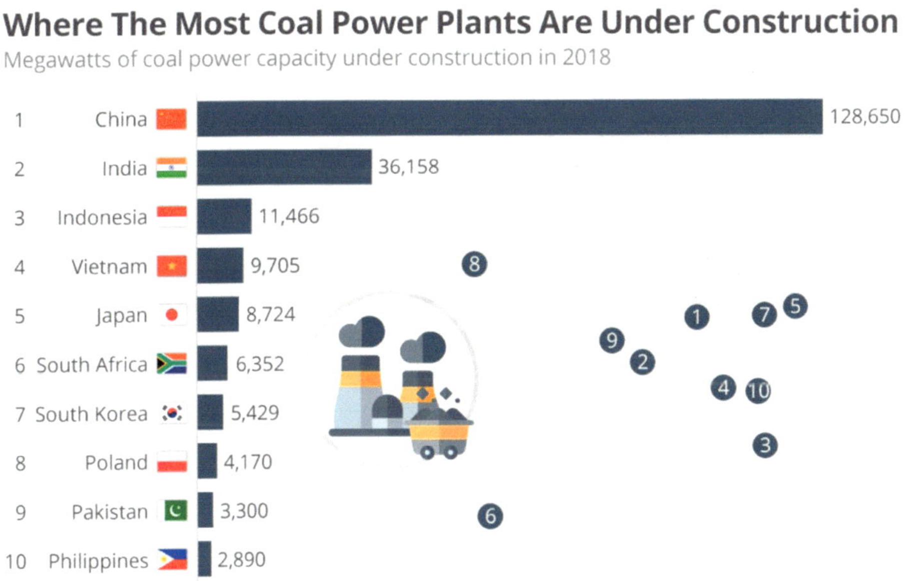

Tom Steyer was the other candidate. A green philanthropist and former Hedge Fund manager who runs the *NextGen Climate Action Committee* which supports candidates who embrace climate change as a major issue. He is the largest individual political donor of the past decade, having handed over hundreds of millions of dollars to influence Presidential and State elections.

Money for Nothing

Vast sums of money are made by the operators and owners of wind and solar farms and not just from the ultra-expensive energy these Green trinkets provide the grid. In 2019 operators of 86 wind farms in Britain were paid more than £136 million in so-called "constraint payments". This occurs when the wind is blowing so hard that the grid cannot cope with the surge. The wind farms are paid to cease producing electricity and actually receive higher sums of compensation than they would have received if they had been producing electricity[14]. The subsidies are paid if the wind farms are unable to operate. How was this scam permitted? Perhaps the Governments that established the scheme should provide the answer. Imagine being paid more for producing no electricity than for providing energy.

It is not just renewables that have been raking in the cash but also electricity suppliers. Owing to the gradual replacement of cheap conventional electricity with expensive renewables, some energy companies have seen their profits triple[15]. Electricity generators make the biggest profits when demand is high because they agree to sell power at highly inflated prices in order to prevent any blackouts.

Other companies have made millions from inept Governments. The Vales Point coal-fired plant in New South Wales was sold for just $1 million in 2015 to a private company, Sunset Power International. It is now worth over $800 million[16].

Why did the State Government sell-off such a valuable asset for less than the price of a small house? Perhaps that question should be directed at the aforementioned Kerry Schott who oversaw the sale. According to Alan Moran of *Regulation Economics*[17], the money trail is obscure when it comes to finding out just who is profiting from the destruction of the formerly cheap and reliable electricity grid but the chief suspects often hide in plain sight.

Money has flowed into all sorts of different areas relating to global warming. After palaeontologist Tim Flannery made his spurious predictions about empty dams and cities running out of water, Australian states enthusiastically built desalination plants at huge expense to the taxpayer. The unions, banks and big business have received massive windfalls whilst the desalination plants have hardly been utilised. Only Perth seems to have benefited from their plant. Sydney's plant was only turned on in 2019, twelve years after construction began. However, the fact is that:

> when the plant is not operational, and just sitting at the ready, it costs $90 a year per house...Now the trigger has been pulled, Sydney residents can expect their water bill to climb by a further $25 to $30 a year.[18]

Desalination also consumes vast quantities of electricity. The New South Wales State Government decided to build a hugely expensive wind farm (using taxpayers' money) at Bungendore near Canberra. Again, unions, banks and big business benefit financially from this project, while consumers are fleeced for the highest electricity and water prices ever. Politicians love to spruik renewable projects claiming they will create jobs; there are just fifteen (15) ongoing jobs at the Bungendore wind farm.[19]

Then there are the schemes such as the Pink Batts Program where the Labor Government of 2007-10 splashed money on installing home insulation, ostensibly to combat climate change. Four people died, hundreds of homes burnt down and when the bungled scheme was abruptly ended, many small businesses went broke costing billions of dollars.

Money is also given to countless failed renewable energy schemes. The Rudd Government was persuaded by their Chief Climate Scientist, Tim Flannery, to hand over $90 million of taxpayers' money to the Geodynamics geothermal project in which Flannery was a shareholder[20]. In the U.S. the Solyndra solar company scandal, a signature piece of the Obama administration, saw over half a billion dollars go up in smoke when the company filed for bankruptcy.

Then there was Crescent Dunes, the Nevada molten salt solar-energy plant with a $737 million Government loan guarantee[21], it is now also bankrupt having produced hardly any power and created only forty-five permanent jobs for the few short years it was operating. Fortunately, a similar project in Port Augusta, South Australia was canned before a further $650 million was wasted. Tens of millions of taxpayers money has also been handed over to failed wave power projects[22] which have been abandoned to rust in the sea[23].

Odd Jobs for Goldfingers

The people who push these projects often end up joining the Boards or getting positions within the companies and then get paid hundreds of thousands even millions of dollars for their "expertise". Or they cash in on shares, buying and selling at just the right time and profiting handsomely just before the scheme collapses and the public are left to clean up the mess, physically and financially. In the private sector numerous companies have been held to account for financial malfeasance, fraud, misleading shareholders and deception. Enron,

WorldCom, Bernie Madoff and, more recently, Theranos are all well-known examples. However, in the global warming industry nobody is ever asked to pay back the money; nobody has ever been arrested let alone investigated. *Au contraire*, the same people and companies are allowed to rort, profiteer and carpetbag over-and-over again with zero accountability and zero responsibility.

Much of the profiteering from global warming occurs in trading and finance, sectors that are largely obfuscated from the public. Whilst you might have heard terms such as carbon trading, emissions trading, cap-and-trade schemes, carbon credits and offsets, the mechanisms and costs are vague and nebulous, often deliberately so. These carbon credit and trading schemes that have been established globally are a rort worthy of a James Bond villain. The absurdity that people are trading a naturally occurring invisible gas that makes up a miniscule part of the atmosphere underlines how out of hand things have become. They have literally created a market out of thin air. Who benefits? The Big Banks, traders and financial institutions who often have a nexus with the political arena. For example, the legislation for Australia's carbon trading scheme, the euphemistically named "Safeguard Mechanism", was drafted less than a month after former Goldman Sachs partner Malcolm Turnbull elbowed his way to the Prime Ministership and was enacted the following year.

Further down the bottomless pit of climate change cash, but no less innocent, are the alarmist scientists and their innumerable supporters that live lavishly off taxpayers' money. The simplest and easiest way to deal with man-made global warming would be to switch all funding from finding a link between CO_2 and warming to disproving any link between it. Man-made global warming would vanish overnight.

Scientists who pursue the CAGW hypothesis are guaranteed not only lavish Government grants but also prestige and the luxurious overseas conferences in exotic locations at five-star hotels. Climate science is now more akin to an inverted gameshow where the contestants with the most alarming and dubious answers win millions of dollars and luxury holidays.

Many millions of dollars of funding have been provided on the assumption that the Great Barrier Reef is in decline and more research is required to work out how to rescue it. Typical of the government approach to funding this topic was a grant of $444 million allocated to the Great Barrier Reef Foundation in 2018 on instruction from then Prime Minister Malcolm Turnbull without even submitting a grant application with a plan to use it. The reef is a cash cow for researchers and they milk it by promoting the fake news that the reef is on its last legs. The locals know that the news is fake and so do the boatloads of tourists who leave the docks at Cairns every day and return with vivid memories and pictures of the delights of the coral and the other marine life. If you are not qualified or interested in the reef, then fear not as there are tens of millions of dollars available every year to produce studies on how climate change will mean the end of walnuts[24], penguins hatching smaller chicks[25], tea to taste different[26] and even how saying goodbye at the airport contributes to global warming[27].

The Green Machine

The alarmists were very quick out of the blocks, long before any debate, enacting that infamous adage that you should always accuse the other side of that which you are guilty of yourself. They projected the notion that anyone who opposed their message must be "in the pay of Big Oil or Big Coal". However, this is farcical when one considers the staggering sums of money available to the alarmist side. The accusation is also absurd since companies like Exxon are not just oil, coal and gas producers. They make plenty of money harvesting government grants and subsidies for renewable energy along with many other large corporations like General Electric and British Petroleum. These corporations have become political entrepreneurs rather than market entrepreneurs.

Moreover, the fact is that many sceptics and realists ("deniers") are less likely to be in the pay of anyone. They mainly work without charge, requiring no recompense and in their own spare time. They are not attracted by fame, glory or any sort of freebies. They are motivated by sincere convictions and because they hold the scientific method in high regard, prefer truth over fiction and can see the catastrophic damage caused by global warming alarmism.

The Global Warming industry is a leviathan and its lobbyists are connected at every level with Big Government. When Government pays for Government institutions, increasing their financial and institutional power, it is hardly surprising that these organisations will always be biased in their own favour. No Government funded institution has ever voluntarily shut itself down. Their entire *raison d'etre* is to invent, find, inflate and conflate problems that need to be solved, usually by more money and more power for that agency or institution.

A minority are profiting and profiteering from the colossal climate change cash bonanza. This minority includes corporations, Big Banks, those in politics, academia all the way down to those who can afford solar panels. But it is the 99% who are losing out. We are compelled at every level to provide the money to finance the elite alarmists, whether it is directly through higher power bills, indirectly through taxation or hidden costs. These hidden costs occur when industry must comply with rules on carbon emissions and the cost of so doing is passed on to consumers.

If you do not like coal or oil then you can eschew the car and buy a pushbike, stop flying and switch off the mains but there is no choice when it comes to funding renewables, paying alarmist scientists, providing money for research and bearing the costs of their inducements. The money is taken out of our pockets without our consent and often without our knowledge. The unfortunate truth is that, despite all the trillions of dollars already spent, there has not been a single improvement; we are told things are getting worse and worse which will, of course, require more and more money to "fix".

The Green Machine is a self-perpetuating viral monster that will just grow and grow and the only thing its proponents are interested in "greening" is their own pockets. One estimate paints a bleak future, not for the environment but our financial well-being, with a potential seven trillion dollars of "investment opportunities" in the global carbon reduction market[28]; that is nearly 5 pages of the *Annual Spending on Climate Change* diagram. These "investment opportunities" are simply the transference of vast sums of wealth and power from working people to the financial and political elites. Global warming is not the greatest catastrophe facing mankind, it is the greatest heist in human history.

Afterword

"It was the worst of times...it was the age of foolishness...it was the epoch of incredulity...it was the season of Darkness...it was the winter of despair...we had nothing before us...we were all going direct [to Hell]."

EDITED EXCERPT FROM *A TALE OF TWO CITIES*, CHARLES DICKENS

You would be forgiven for thinking the above sounds like something written by a climate alarmist or hissed from the mouth of an overindulged Nordic teenager. It more accurately describes how rationalists, realists, firm believers in truth, evidence and the scientific method have felt about the Global Warming debate in the past couple of decades.

Like the xenomorph from the 1979 movie *Alien* we have watched the small, mewling, malformed Global Warming freak burst forth in the last decade of the 20th Century and then grow at a horrifying rate into the huge devouring Climate Change monster that we just cannot escape from. It pervades every aspect of our lives; the media suffocates us with it, activists demand we divest and boycott, the education system indoctrinates children so not even the dinner table is free from it, everything we do or buy is labelled with some environmental virtue-signalling messaging with an added dose of guilt-shaming. And then there are the awkward conversations with friends and colleagues when they discover you are a "denier".

It is impossible to find anyone today who does not hold an, often very polarised, opinion on global warming. Over the years of doing battle a stark and noticeable difference between the two sides has become evident: Realists debate, Alarmists orate. And then there are those who are stuck in the middle who, just to fit in, often feel the need to regurgitate the fiery exhortations from the tyranny of "experts" and their media darlings.

Today everyone believes they are an expert because they watch the ABC or CNN. Yet when you actually question the knowledge of many alarmists and fitters-in, it turns out they have little idea of what they're talking about. One of the principal motivations for writing this book was that when I asked some fairly simple questions such as "What is the current percentage of CO_2 in the atmosphere?", "What is the main greenhouse gas?", "How old is the Earth?", "Are there more trees today than 100 years ago?" or "How much of the Antarctic has melted?" the answers were jaw-dropping.

Approximately 97% could not answer a single question correctly. Often, they were embarrassingly wrong; CO_2 was definitely the main, sometimes only, greenhouse gas and comprised anything from 5% all the way up to 50% of the atmosphere, half the Antarctic had melted which was killing the polar bears and walruses as they fell into the sea, CO_2 pollution would block out the sun and (inexplicably) cause the Earth to heat up, all the plants and trees would die out and the corals would be dissolved in acid.

Very few admitted they did not know the answers. The even smaller group that answered correctly were almost all over 50 years old. You know, those people who went to school when we had an education system that actually educated. A small hint: if you are one of the 97%; it is better to admit you don't know than to guess and make a fool of yourself.

Whilst the useful idiots and foot-soldiers of the CAGW movement are aggravating and child-like in their approach we have to remember they get their very limited knowledge of CAGW from only a few sources; a very partisan media populated by ignoramuses, a disastrously abysmal education system long goose-stepped through by the left, and relentless peer pressure from social media coupled with the desire not to be the nail that sticks out, an issue especially crucial to school children.

It is amazing and disheartening how many people take the word of pundits, politicians and media 'presstitutes' without question and without thought. They are merely regurgitating the sermons and preaching of others with no critical thought or introspection. They are incapable of answering or understanding the basic precepts of what they are regurgitating. They are effectively members of a cult. Cults can do crazy things: the Taiping Rebellion, Heavens Gate, The Peoples Temple and Aum Shinrikyo spring to mind as just a few examples. By contrast, contemporary Global Warming cultists glue themselves to roads, weep for trees and decide not to have children (not necessarily a bad thing!). Yet the damage goes far beyond blocked roads and children skipping school to protest. It is ordinary people, the ones the elites hold in patronising contempt, that will ultimately pay the heaviest price: the financial burdens, the lost jobs and opportunities, the lower standards of living and eventually fewer years on the planet. It is for those that this book is dedicated to, lest they soon be living in a bleak house in hard times.

It is curiosity, a desire to know, a reverence for the scientific method and faith in truth that you need to set out on the journey back to reality. Congratulations and thank you for taking a few steps on that journey by reading this book.

REFERENCES

Chapter I

[1] The Nexus of Biofuels, Climate Change, and Human Health: Workshop Summary. Roundtable on Environmental Health Sciences, Research, and Medicine; Board on Population Health and Public Health Practice; Institute of Medicine. Washington (DC): National Academies Press (US); 2014 April

[2] Ethanol plant. Licensed from Adobe Stock Standard Licence

[3] Rice, T, "Biofuels are driving food prices higher", *The Guardian*, 1 June, 2011.

Runge, C F and Senauer, B, "How biofuels could starve the poor", *The New York Times*, 7 May, 2007.

Format IOM (Institute of Medicine), "Sustainable diets: Food for healthy people and a healthy planet." Workshop summary, Biofuels and Food Security Issues, The National Academies Press, Washington DC. 2014.

[4] Hulme, M, *Why We Disagree About Climate Change: Understanding Controversy, Inaction and Opportunity*, Cambridge University Press, Cambridge, 2009.

[5] Fact Sheets, *Household Air Pollution and Health*, WHO Newsroom, 6 March, 2018.

[6] Naidoo, K, "Renewable energy for all: How an Indian village was electrified", *Greenpeace*, 27 April, 2015.

Jha, G, "Indian village demands real electricity, not solar gimmick", *The Global Warming Policy Forum*, 27 August, 2014.

Vaidyanathan, G, "Coal trumps solar in India", *Scientific American*, 19 October, 2014.

[7] Bill Leak, *The Australian*, 14 December, 2015. Quote from Paul Whittaker, editor of *The Australian*.

[8] Grover, D "What is the demographic transition model?", *Population Education*, 13 October, 2014.

[9] Martin, D, "23bn goes up in smoke! Chris Huhne's pet scheme to fit 'green' energy boilers is open to fraud and makes pollution worse", *Daily Mail*, 17 May, 2018.

[10] Wikipedia, "Renewable heat inventive scandal", Wikipedia, accessed 27 September, 2020.

[11] Watts, A, Nobody expects the Spanish solar inquisition!, August 17, 2014.

[12] Carbon Brief, Mapped: Climate change laws around the world, May, 2017.

[13] Shining a light: Australia's gas and electricity affordability problem, National Press Club, 20 September 2017.

[14] Butler, B, "Australian aluminium smelters 'not sustainable' due to high power costs, Rio Tinto says", *The Guardian*, 1 November, 2019.

[15] Moran, A, *The Hidden Cost of Climate Policies and Renewals*, Regulation Economics, 2020.

[16] United States. Congress. Senate, Costs of Inaction: The Economic and Budgetary Consequences of Climate Change, Hearing Before the Senate Budget Committee, One Hundred and Thirteenth Congress, Second Session, July 29, 2014

[17] Deighton, L, "Climate -change contrarian loses Australian funding," *Science*, 22 October, 2015.

Chapter II

[1] Editorial, 2014, "Google Scientists Admit Renewable Energy Can't Work," *Investor's Business Daily*, 24 November.

[2] Franz Lawaszeck, "Technology and Economy in Third Reich: A Program for Work", published in 1933.

[3] Solect Energy, 'The History Of Solar Energy: Let's Geek Out', Accessed: 24th June, 2020.
Paraskova, T, "The Biggest Proponent of Solar Energy Isn't What You'd Expect," *Oil Price Com*, 2 October, 2019.

[4] Commentary, "Big Wind's Dirty Little Secret: Toxic Lakes and Radioactive Waste, *Institute for Energy Research*, 23 October, 2013.

[5] News, "Land Needs for Wind, Solar Dwarf Nuclear Plant's Footprint," *Nuclear Energy Institute*, 9 July, 2015.

[6] Airborne Wind Energy: Advances in Technology Development and Research, edited by Roland Schmehl, Springer, 2013, pg.7.

[7] Mackay, D J C, "Sustainable Energy Without the hot Air," UIT, Cambridge, 2008.

[8] Sahagen, L, "This Mojave Desert solar plant kills 6,000 birds a year," *Los Angeles Times*, 2 September, 2016.

[9] SRSrocco Report, "The Renewable Green Energy Myth: 50,000 Tons of Non-Recyclable Wind Turbine Blades Dumped In The Landfill", 2020.
[10] American Elephants, "14,000 Abandoned Wind Turbines Litter the Unites States", July 7, 2013. Image: Benjamin Rasmussen / Getty Images.
[11] The Environmental Costs of Renewable Energy Part 1, J Grimshaw 2021 ©.
[12] Jones, B, "Child Miners Aged Four Living a Hell on Earth so You Can Drive an Electric Car," *The Mail on Sunday*, 6 August 2017.
[13] The Environmental Costs of Renewable Energy Part 2, J Grimshaw 2021 ©.
[14] Coal-seam gas workers. AP Photo / Ralph Wilson & Julien Harneis, Child labor, Artisan Mining in Kailo Congo, https://www.flickr.com/photos/16935515@N00/1873057946 Licence: https://creativecommons.org/licenses/by-sa/2.0/deed.en
[15] Quirk, T, "Problems and Limits for Wind Power," *Jo Nova blog*, October 12, 2016.
[16] Skinner, B, "The RERT Locker", *Australian Energy Council*, 2 March, 2018.
[17] Potter, B, Macdonald-Smith, A & Evans S, "Alcoa Smelter Ordered to Power Down as AEMO Takes Action Amid Victoria, SA Heat," *Financial Review* 25 January, 2019.
[18] Harrison, D, "What Caused the Blackouts in Melbourne, and do Victorians Need to Get Used to Power Cuts?" *ABC News*, 25 January, 2019.

Chapter III

[1] *The Energy of the Future: Sixth "Energy Transition": Monitoring Report.* (Berlin: Federal Ministry for Economic Affairs and Energy, 2008).
[2] Wetzel, D, "Germany risks permanent loss of control of energiewende, federal audit office warns," *Die Welte*, 29 September, 2018.
[3] "Nuclear Power Phase-Out," *Wikipedia*. Accessed 24 September 2020.
[4] "'Renewables are cheap' myth busted: Full cost of wind & solar simply staggering," 2018, *Stop These Things*, 1 April.
[5] Sales, L, "Rudd on the green offensive", Australian Broadcasting Corporation Tue 30 October, 2007.
[6] "Extreme green poverty: Germans face E44 billion bill for electricity alone in 2019," *The Global Warming Policy Forum* 29 November, 2019.
[7] Sloan, J, "State surge of unaffordable energy measures", The Australian, June 10, 2017.
[8] Energiewende 2018, p9.
[9] Rueb, E S, "How New York gets its electricity", *New York Times*, 1 October, 2017.
[10] Efstathiou, J, "New York's Plan to Kill Coal Is Already Boosting Power Prices", Bloombert News, 18 January, 2019.
[11] Rueb, E S, "How New York gets its electricity", *New York Times*, 1 October, 2017.
[12] Northey, H, "No matter who's dealing, DEO pick usually a 'wild card'," *E&E News*, 12 August, 2016.
[13] Parkinson, G, "Cost-effective, clean, reliable grid: 'you can have it all' says Zibelman", *Renew Economy*, 22 February, 2019.

Chapter IV

[1] Aerial view of rainforest. Licensed from Adobe Stock Standard Licence
[2] The Global Thermostat. J Grimshaw 2021 ©
[3] Gabbatiss, J, "Global warming drove 'explosion' of life 500 million years ago that ultimately gave rise to humans", *Independent*, 9 May, 2018.
[4] "Polar ice caps and geologic history", *Science Encyclopedia of Science and Philosophy*, accessed 2 October, 2020.
[5] Patowary, K, "The frost fairs of river Thames", *Amusing Planet*, 7 April, 2018.
[6] Mokyr, J, "Great famine, Ireland 1845-1849", *Britannica on Line*, accessed 25 September, 2020.
[7] Kininmonth, W, "Dear Mr Abbott", *Quadrant Online*, 20 May, 2010.
[8] Chen, S, "The climate theory casting new light on the history of Chinese civilization", *South China Morning Post*, 24 September, 2019.
[9] Smith, P, "UK has sixth-highest rate of excess winter deaths in Europe", *E3G*, 23 February, 2018. Coyle, H, "48,000 Brits dead after worst winter in 42 years", *The Daily Star*, 7 April, 2018.

Agerholm, H, "Dozens die across Europe as cold weather blights continent", *The Independent,* 2 March, 2018.

Chapter V

[1] The temperature reconstruction is by paleogeographer C.R. Scotese and the CO_2 reconstruction is by R.A. Berner & Z. Kothavala, "Geocarb III: A revised model of atmospheric CO_2 over Phanerozoic time." *American Journal of Science*, Vol. 301, Feb. 2001, pp. 182-204.
[2] Atmospheric CO_2 and Global Temperature, Prof. Ole Humlum, *Climate4you,* August 2013.
[3] The Hockey Stick Graph, Report of Working Group I of the Intergovernmental Panel on Climate Change, Summary for Policymakers, Fig. 2.20, p.3, 2001.
[4] Peck, S, "Al Gore's 'Nine Inconvenient Untruths'," *The Telegraph,* 11 October, 2007.
[5] Petit, Barkov et al., "Climate and atmospheric history of the past 420,000 years from the Vostok ice core, Antarctica", *Nature*, Vol. 399, 3 June, 1999 & "The 800 year lag in CO_2 after temperature – graphed", joannenova.com.au. Accessed August 2019.
[6] "NASA Research Finds 2010 Tied for Warmest Year on Record", NASA GISS, January 12, 2011.
[7] Temperature curve last fifteen thousand years from central Greenland ice core. Adapted from " The Climate of the Holocene", University of Arizona, 2016.
[8] Arrhenius, S, *Worlds in the Making: The Evolution of the Universe*, Harper & Brothers, London: p. 63, 1908.
[9] Abbot, J and Nicol, J, "The Contribution of Carbon Dioxide to Global Warming," *Climate Change: The Facts 2017*, IPA, Melbourne.
[10] The Atmosphere in Stadium, J Grimshaw 2021 ©.
[11] Ma, Q, "Greenhouse Gases: Refining the Role of Carbon Dioxide," *NASA Science Briefs*, March, 1998.
[12] CO_2 Parts Per Million, J Grimshaw 2021 ©.
[13] Donohue, R, "Deserts 'greening' from rising CO_2," *CSIRO News releases and briefings*, 3 July, 2013.
[14] The Carbon Dioxide Dial. J Grimshaw 2021 ©.
[15] Voigt, K, "Quake moved Japan coast 8 feet, shifted earth's axis," *CNN*, 20 April, 2011.

Chapter VI

[1] Odenwald, S, Dr., "The Day the Sun Brought Darkness", *NASA* website, March 14, 2009.
[2] Klein, C, "A Perfect Solar Superstorm: The 1859 Carrington Event", *The History Channel*, March 14, 2012.
[3] Mechanism of the solar system planets. Licensed from Adobe Stock #346824001.
[4] Earth's Tilt Wobble and Orbit, J Grimshaw 2021 ©.
[5] Mathez, EA, "Milutin Milankovitch: Seeking the Cause of the Ice Ages", *American Museum of Natural History* website, 2000.
[6] Buis, A, "Milankovitch (Orbital) Cycles and Their Role in Earth's Climate", *NASA's Jet Propulsion Laboratory*, February 27, 2020.
[7] "How much water is in the ocean?", *National Ocean Service* website, 2020.
[8] Gabbatiss, J, "Active volcano discovered beneath Antarctic ice sheet could be contributing to rapidly melting glacier", *The Independent,* 29 June, 2018.
[9] Howard, BC, "New Giant Volcano Below Sea Is Largest in the World", *National Geographic*, September 7, 2013.
[10] Oskin, B, "Scientists Link Underwater Eruptions to Climate Change", *NBC News*, February 7, 2015.
[11] El Niño & Other Oscillations", *Woods Hole Oceanographic Institution* website, 2020.
[12] Plimer, I, *How to Get Expelled From School*, Connor Court, Australia, 2012, p. 51
[13] Layers of the Atmosphere, J Grimshaw 2021 ©.

Chapter VII

[1] Unstoppable Solar Cycles: The Real Story of Greenland, YouTube Video, Accessed: 24th October, 2020.
[2] Seitz, Frederick, "A Major Deception on Global Warming", *The Wall Street Journal*, June 12, 1996.

[3] Pielke, R, "My 1995 Resignation Letter From The IPCC," *Climate Science Reporting*, September 30, 1995.
[4] Recent News on Energy, "Resignations and Withdrawals From the IPCC", *Habitat 21*, Accessed: 17 November, 2019.
[5] Rose, D, "Green 'Smear Campaign' Against Professor Who Dared to Disown 'Sexed Up' UN Climate Dossier," *The Mail on Sunday*, February 20, 2015.
[6] *NOconsensus.org*, 2010, "UN's Climate Bible Gets 21 'F's on Report Card."
[7] Laframboise, D, 2011, *The Delinquent Teenager Who Was Mistaken For the World's Top Climate Experts*, Createspace Independent Publishing Platform, p.33
[8] Laframboise, *ibid* p. 41.
[9] Laframboise, *ibid* p. 48.
[10] Laframboise, D, "Meet the IPPC's Youngest Lead Author," *No Fracking Consensus.com,* October 21, 2010.
[11] Staff Profile, "A/Prof Lisa Alexander," *Australian Research Council*, 2017.
[12] Botkin, D B, "Science and Soothsaying", *International Herald Tribune*, 28 December, 2007.
[13] Schiermeier, Q, "IPCC Flooded by Criticism," *Nature* 463, pp.596, 2 February, 2010.
[14] About the NIPCC The Nongovernmental International Panel on Climate Change. *The Heartland Institute.* Accessed 30 November 2020.

Chapter VIII

[1] IPCC third assessment Report; Section 14.2.2.2, p.774, 2001.
[2] Spencer, R, Dr., "The Five Questions Global Warming Policy Must Answer", Roy Spencer Blog. Accessed, December 2019.
[3] Matotaka, M, *Confessions of a Climate Scientist. The Global Warming Hypothesis is an Unproven Hypothesis*, Kindle, 2019.
[4] Screenshots from Rockstar's Grand Theft Auto V and Microsoft's Minecraft.
[5] Gosselin, P, "Earth Surface Temperature too Scant, No Certainty. Southern Hemisphere Stations Show Cooling Since 1880", *NoTricksZone,* August 23, 2019.
[6] Marohasy, J, "Temperature Change at Rutherglen in South-East Australia", *New Climate,* 2016.
[7] The Australian Academy of Science, *The Science of Climate Change: Questions and Answers*, 2015.
[8] Models pose for a group photograph during a photocall before the Victoria's Secret Fashion Show at the Grand Palais in Paris. Licensed from Adobe Stock #152011750.

Chapter IX

[1] Shapley, D, "The Genesis Strategy", *The New York Times* July 18, 1976.
[2] Stein, R, Fahl K, Schade I, Manerung A, Wassmuth S, Niessen F, "Holocene Variability in Sea Ice Cover, Primary Production and Pacific Water Inflow and Climate Change in the Chakchin and East Siberian Seas (Arctic Ocean)", *Journal of Quaternary Science* Vol 32 (3): 362-379, 2017.
[3] Agency report, "Arctic ice 'could be gone in five years'", *The Telegraph* 12 December, 2007.
[4] Vidal, J, "Arctic expert predicts final collapse of sea ice within four years", *The Guardian*, 17 September, 2014.
[5] Amos, J, "Arctic summers ice-free 'by 2013' ", *BBC News,* 12 December, 2007.
[6] Commentary, "Arctic ice could be gone by 2030", *The Telegraph*, 16 September, 2010.
[7] Harden, B, "Experts predict polar bear decline", *The Washington Post*, 7 July, 2005.
[8] Goode, E, "Human-driven global warming is biggest threat to polar bears, report says", *The New York Times*, 9 January, 2017.
[9] Crockford, S, *Ten Good Reasons Not To Worry About Polar Bears*, The Global Warming Policy Foundation, 2013.
[10] LAST CHANCES:
Bonn 2001. "A Global Warming Treaty's Last Chance. That teetering edifice that is the Kyoto Protocol gets some emergency repair work this week as delegates from 180 countries gather in Bonn", *Time Magazine*, 16 July, 2001.

Montreal 2005. "In an open letter to delegates at the Montreal environmental summit, 'I'm scared. For 15 years I've watched international progress on climate change get slower and slower, even while the pace of global warming seems to get ever more rapid. With time running out for the global climate, your meeting in Montreal represents a last chance for action.'" *The Independent*, 28 November, 2005.
Bali 2007. "World leaders will converge on Bali today for the start of negotiations which experts say could be the last chance to save the Earth from catastrophic climate change", *The New Zealand Herald*, 3 December, 2007.
Poznan Poland, 2008. "The world will "suicide" if it cannot strike a strong climate pact soon, Australian environmental scientist Tim Flannery has warned. 'This round of negotiations is likely to be our last chance as a species to deal with the problem.'" *The Age*, 9 December 2008.
Copenhagen 2009. "The Copenhagen summit is the world's last chance to save the planet from "catastrophic" global warming, according to a major study led by Lord Stern of Brentford, the country's leading authority on climate change." *The Telegraph*, 2 December 2009.
Cancun 2010. "A sense of foreboding is one of the few points of general agreement among the 15,000 participants congregating for the next two weeks. Jairem Ramesh, the Indian environment minister, sees it as the "last chance" for climate change talks to succeed.", *The Telegraph*, 29 November, 2010.
Durban 2011. "Rev. Dr. Olav Fyske Tveit, who leads the World Council of Churches, says the upcoming climate conference in South Africa is mankind's 'last opportunity' to address climate change.", *Spero News*, 27 November 2011.
Durban 2011. "Durban climate change meeting is "the last chance". Attended by over 200 countries, this week's major UN conference has been described by many experts as humanity's last chance to avert the disastrous effects of climate change.", *UCA News*, 28 November, 2011.
Doha 2012. "Tomorrow, the whole world talks about irreversible global warming as this year's international climate change summit begins. Participating are 195 countries (almost all of the United Nations).", *The Examiner*, 25 November 2012.
Warsaw 2013. "Is the Warsaw Climate Change Conference a last-chance summit? The Warsaw Climate Change Conference opened on Monday 11th November. After the 2012 failure of Doha, this summit could represent a turning point in the fight against global warming.", *Sustainable Mobility*, 14 November, 2013.
Lima 2014. "WWF issued the following statement today from Samantha Smith, Leader of WWF's Global Climate and Energy Initiative, as the UN climate talks drew to a conclusion: 'A repeat performance next year would be disastrous,'" *WWF Global*, 23 November, 2013.
Paris 2015. "Scientists are calling on world leaders to sign up to an eight-point plan of action at landmark talks in Paris. The key element is the goal to limit global warming to below 2C by moving to zero carbon emissions by 2050. The UN meeting in December is "the last chance" to avert dangerous climate change, according to the Earth League.", *BBC News*, 22 April, 2015.
[11] Ehrlich, P Dr., The Population Bomb, Sierra Club-Ballantine Books, 1968, p. xi.
[12] Blair, T, "The Great Die-Off", *The Daily Telegraph*, April, 2016.
[13] Tracinski, R, "Seven big failed environmentalist predictions", *The Federalist*, 24 April, 2015.
[14] Brand, S, "Rethinking extinction", *Aeon* 21 April, 2015.
[15] *Ibid.*
[16] Stevens, W K, "Scientists say earth's warming could set off wide disruption", *The New York Times*, 18 September, 1995.
[17] Seenan, G, "Global warming forces sale of Scottish winter sports resorts", *The Guardian*, 14 February, 2004.
[18] Onians, C, "Snowfalls are now just a thing of the past", *The Independent*, 20 March, 2000.
The Independent has removed this article from the internet, only screenshots exist on a few sites.
[19] Bennet, T, "All classes at UEA have been cancelled because of the snow", *The Norwich Tab*, 2 March, 2018.
[20] Global Snow and Ice – Annual, *NOAA National Center for Environmental Information*, 21 January, 2013.
[21] Sinpetru, L, "Global Warming Caused the Cold Snap, the Obama Administrations Says", Softpedia, 9 January, 2014.

[22] Williams, R, "Reporting CO_2", *ABC TV* 10 March, 2007.
[23] "Queensland Floods", Wikipedia, accessed 23 September, 2020.
[24] Inman, M, "Strongest hurricanes may double in frequency, study says", *National Geographic News*, 22 January, 2010.
[25] IPCC's Fifth Assessment Report, 2014.
[26] NASA/Goddard Space Flight Center, "NASA detects drop in global fires", ScienceDaily, accessed September 23, 2020.
[27] OFDA/CRED International Disaster Data / Bjorn Lomborg.
[28] "Increases in air temperature can promote wind-driven dispersal and spread of plants", *Proc. R. Soc. B* published online before print June 10, 2009.
[29] Morano, M, 2009, "AP's Borenstein Reports: Winds 'seems to be dying down...the cause may be global warming'", *Climate Depot*, 6 November, 2009.
[30] Ahmed, S, "Tea and the taste of climate change", *Issue*, 103: 44-51, July, 2014.

Chapter X

[1] Cook, J, Nuccitelli, D, Green, S A, Richardson, M, Winkler, B, Painting, R, Way, R, Jacobs, P and Skuce, A, "Quantifying the consensus on anthropogenic warming in the scientific literature," *Environmental Research Letters*, volume 8 (2), 2013.
[2] Oreskes, N, "The Scientific Consensus on Climate Change", Science Vol 306, Issue 5702, 3 December, 2004.
[3] *Ibid.*
[4] Doyle, A, "Scientists say united view on global warming, at odds with public view," *Reuters*, 16 May, 2013.
[5] Cook *et al*, *ibid.*
[6] Cook *et al*, *ibid.*
[7] Obama, B, Twitter, May 17, 2013.
[8] Tuttle, I, "The 97 percent solution," *National Review*, 8 October, 2015.
[9] Science By Consensus - Geocentrism. J Grimshaw 2021 ©.
[10] Science By Consensus - Smoking. J Grimshaw 2021 ©.
[11] Crichton, M, "Aliens cause global warming", Caltech Michelin Lecture, January 17, 2003.
[12] Global Warming Petition Project, 2020, accessed 30 September 2020.
[13] Clark, T D, "Ocean acidification does not impair the behaviour of coral reef fishes," *Nature*, **577**: 370-375, 2020.
[14] Schiermeier, Q, "Investigation finds Swedish scientists committed scientific misconduct", *Nature*, 7 December, 2017.
[15] Richardson, V, "Calls to punish skeptics rise with links to climate change, hurricanes," *The Washington Times*, 11 September, 2017.

Chapter XI

[1] Galileo Quote, J Grimshaw, 2020. Galileo illustration licensed from Adobe Stock.
[2] Prisk, J, "Misleading smoke gets in the public's eyes", *The Sydney Morning Herald*, 2 May, 2012.
[3] Pocerady coal fired power station, Czech Republic. Licensed from Adobe Stock.
[4] Carington, D, "Why the Guardian is changing the language it uses about the environment", *The Guardian*, 17 May, 2019.
[5] Thunberg, G, "# ClimateBreakdown," *Twitter for iphone*, 5 May, 2019.
[6] Diamonds. Licensed from Adobe Stock
[7] Warming Twice As Fast (WTF!). Collage of headlines, J Grimshaw, 2020.
[8] "How do we connect young people to their environment?", Sustainable Schools NSW website, accessed 5 September 2020.
[9] Burney, M, "In New Jersey schools, climate change education will be mandatory" *Philadelphia Inquirer*, July 3, 2020.
[10] "Global warming in popular culture", *Wikipedia*, accessed 2 October, 2020.
[11] Hymowitz, K S, "Ricky Gervais read America", *City Journal*, 7 January, 2020.

Chapter XII

[1] Harper, J, "Climate change industry' now a $1.5 trillion global business", *The Washington Times*, Tuesday, August 11, 2015.
[2] Driesen, P, "Climate Crisis, Inc", *Town Hall*, 22 August, 2015.
[3] Annual Spending on Climate Change, J Grimshaw 2021 ©.
[4] Knapp, A, "Apollo 11's Anniversary: The facts and figures behind the $152 billion moon landing", *Forbes*, 20 July, 2019.
[5] "ExxonMobil to Join Stanford Strategic Energy Alliance", *ExxonMobil Newsroom* 12 March, 2019.
[6] McCarron, S, "Flawed study claiming ExxonMobil mislead public is disappointing," *ExxonMobil Energy Factor Perspectives Blog*, 1 March, 2018.
[7] Nisbett, M C, "Strategic philanthropy in the post-Cap-and-Trade years: Reviewing U.S. climate and energy foundation funding," *Wires Climate Change*, Volume 9 (4) 22 May, 2018.
[8] Influence Watch Website, Sustainable Markets Foundation, Accessed: 14 December 2019.
[9] Anders, G, "Perfect enough: Carly Fiorina and the reinvention of Hewlett-Packard," Portfolio, New York, 2003, p 109.
[10] Bastach, M, "Climate change 'philanthropy' efforts include $20 million to Al Gore's group," *Daily Caller*, 22 May, 2018.
[11] Markay, L, "Pelosi's husband invested in solar firm weeks before lucrative expansion," *The Washington Free Beacon*, 19 April, 2016.
[12] "Michael R Bloomberg commits $50 million to international efforts to move beyond coal, reinforce leadership on global climate action," *Bloomberg Philanthropies*, 9 November, 2017.
[13] McCarthy, N, "Where the most coal plants are under construction," Statistica, 29 March, 2019.
[14] Malnick, E, "Wind farms paid up to 3 million pounds per day to switch off turbines," The Telegraph, 19 January, 2020.
[15] Pash, C, "AGL has tripled profit while electricity prices soar," *Money & Markets*, 9 August, 2018.
[16] Nova, J, "40 year old coal plant sold for $1M makes $100M profit and will run another 30 years," Jo Nova blog, 7 September, 2018.
[17] A. Moran, Regulation Economics
[18] Cockburn, P, "Sydney desal plant is turned on – so what does this mean?" *ABC News*, 29 January, 2019.
[19] "Capital Wind Farm," *Wikipedia*, accessed 27/9/2020.
[20] Fedorowytsch, T, "Geothermal power project closes in SA as technology deemed not financially viable", *ABC News*, Tuesday 30 August, 2016.
[21] Crescent Dunes, Loans Program Office, US Dept of Energy, September 2011.
[22] Carnegie Clean Energy, Albany, WA. Energetech/Oceanlinx, Port Kembla NSW & Port McDonnell, SA. BPS BioWave, Port Fairy, VIC.
[23] Humphries, G, "$7m handout as Port Kembla eyesore rots," *The Illawarra Mercury*, 12 July, 2013.
[24] Wallheimer, B, "Walnut trees may not be able to withstand climate change," *Science Daily*, Purdue University, 30 November, 2011.
[25] Goswell, G, "Study links Phillip Island penguin breeding patterns to climate change," *ABC News*, 4 September, 2009.
[26] Wei, C, "The future of tea looks bleak, thanks to climate change," *Eater*, 8 July, 2016.
[27] Derbyshire, D, "Save the planet: Taxpayers fund £500,000 research to help you say virtual airport farewell to reduce carbon footprint," *Daily Mail Online*, 23 January, 2010.
[28] Shmuel, J, "Climate change initiatives a $t-trillion funding opportunity for capital markets: Carney," *Financial Post*, 15 July, 2016.

About the Author

Jeffrey R Grimshaw is the son of a teacher and a General Practitioner. He has lived and worked in England, America, Japan and Australia. With interdisciplinary graduate degrees in history and politics, a postgraduate in International Relations from Nottingham University and an MSc in Information Technology, specialising in advanced Artificial Intelligence and predictive modelling, he has a unique insight into the merger of the political and scientific arena that is at the core of the global warming industry.

He has designed and worked with computer models that use historical data to predict and optimise current and future decisions and worked in a wide variety of industries including healthcare, online education, food distribution, chemicals, supply chains and logistics. Jeffrey is also a wordsmith, innovator and a former political candidate.

The author declares no conflicts of interest.

Made in the USA
Las Vegas, NV
03 November 2026

57424181R00079